Side-Channel Analysis of Embedded Systems

Maamar Ouladj • Sylvain Guilley

Side-Channel Analysis of Embedded Systems

An Efficient Algorithmic Approach

 Springer

Maamar Ouladj
Paris 8 University
Paris, France

Sylvain Guilley
Secure-IC, S.A.S.
Paris, France

ISBN 978-3-030-77224-6   ISBN 978-3-030-77222-2   (eBook)
https://doi.org/10.1007/978-3-030-77222-2

This Springer imprint is published by the registered company Springer Nature Switzerland AG
The registered company address is: Gewerbestrasse 11, 6330 Cham, Switzerland

Contents

General Introduction

Today's digital era connects everyone and everything in between through the Internet of Things (IoTs). That situation leads to an omnipresence of embedded systems in our daily life endowed with integrated capability to run cryptographic protocols (consumer electronics, telecommunication and industrial appliances, governmental and military systems, etc.). Those need to manage authorized access, firmware integrity, life-cycle irreversible steps, and so on. Thus, the functional correctness of IoTs is an absolute prerequisite. Nevertheless, unless suitable care is given, cryptographic protocol implementations in both software and hardware happen to leak sensitive information during their execution. Side-Channel Attacks (SCA) is a scientific field whose purpose is the analysis of this leaked information, in order to recover the secret parameters of the protocols. In the seminal paper on Side-Channel Attacks (SCA), published by Kocher *et al.* [1], adversary used the execution time to break the implementation both of Diffie-Hellman, RSA, DSS, and other cryptographic systems [1]. Since this seminal paper, several attacks have been introduced using either the power consumption of the device [2], its electromagnetic/optical/acoustic emanations [3–5], its temperature variations, the running time variations [1,6], or a contention for Central Processing Unit (CPU) resources such as cache memories.

In practice, small devices (such as smart cards) are common targets of SCA. But, also larger devices such as smartphones and computers can be attacked through their side-channel leakage [4,7–9].

The SCA requires two steps, namely:

1. side-channel emanation traces collection from the device;
2. analysis of the traces, altogether referred to as the acquisition campaign, to extract the sensitive information.

The first step requires some acquisition skills, which is typically carried out by automated experimental benches, as, for instance, described in Chap. 3 of [10]. The

M. Ouladj and S. Guilley, *Side-Channel Analysis of Embedded Systems*,
https://doi.org/10.1007/978-3-030-77222-2_1

second step is essentially agnostic with respect to the attacked device, in that it consists in distinguishing the correct secret key from several hypotheses.

As stated in the SCA literature, according to the adversary abilities (knowing or not the leakage model of the device, having or not a copy of this target device, …), the optimal attack is either the Correlation Power Analysis (CPA) [11,12], Mutual Information Analysis (MIA) [13], Linear Regression Analysis (LRA) [14], or the Template Attacks (TA) [15]. The relative merit of these distinguishers has been investigated empirically in some competitions, such as the "DPA contest" [16].

This book addresses the following topics:

- the correct implementation of the SCA distinguishers, and
- their optimal algorithmic implementation in terms of speed.

Instead of carrying out an SCA straightforwardly on the raw traces, one can carry it out using the averaged leakage per message value [17–19]. Leveraging this coalescence one can reduce significantly the SCAs' complexities, while SCAs with and without coalescence are asymptotically equivalent (which is proven in the current book, for the first time). Using the Walsh-Hadamard Transform (WHT), one can speed up the Correlation Power Analysis (CPA) from a quadratic to a quasi-linear complexity, under certain conditions [20]. In this book, we first generalize the use of the Walsh-Hadamard transform for any situation. Consequently, using both of the coalescence principle, the spectral approach, and others optimizations, we provide several improvements of the implementation complexity in both of CPA, LRA, and the Template attacks. All these improvements of the complexities are conflated and put in perspective in this book. They lead asymptotically to optimal SCA assaults, from a computational standpoint, even in presence of countermeasures.

The attacks will be illustrated throughout this book on the running example of the Advanced Encryption Standard (AES [21,22]).

Outline and Contributions

The present book is structured in three parts. The first part consists in the state of the art about known attacks. The second part contains the efficient processing of the prominent attacks put forward in the first part.

The beginning of the Part I provides a state of the art of Side-Channel Analyses (SCAs) through three chapters. Chapter 2 consists in some foundations of side-channel attacks. Namely, a general framework of SCA is introduced, then different leakage models are given, and eventually different security metrics are defined. Also we introduce different techniques used for trace pre-processing, leakage-wise.

Chapter 3 tackles the most important side-channel distinguishers. Namely, we introduce Simple Power Analysis (SPA), Differential Power Analysis (DPA), Correlation Power Analysis (CPA, either using Pearson or rank-based statistics), Covariance-Based Distinguisher, Collision Side-Channel Attacks, Mutual Information Analysis (MIA), Kolmogorov-Smirnov Distance (KS)-based distinguisher, Chi-squared Test

(Chi-2-Test)-based distinguisher, Template Attack (TA), Linear Regression-based Side-Channel Attacks (LRA), Machine Learning Distinguishers, and the most recent Higher Order Optimal Distinguisher (HOOD).

In Chap. 4, we introduce the different countermeasures aiming at stymying SCA attacks, namely, the hiding (on the time/amplitude dimension) and the masking techniques. For the time dimension's hiding, we introduce both of the random insertion of dummy operations, the jittering, and the shuffling of operations. For the amplitude dimension's hiding, we introduce the two complementary techniques for reducing the Signal-to-Noise Ratio (SNR), namely, the noise increase and the signal reduction. For the masking countermeasure, we introduce different known masking schemes, namely, the Boolean Masking (BM), the Multiplicative Masking (MM), the Affine Masking (AfM), the Arithmetic Masking (ArM), the Polynomials-based Masking (PM), Leakage Squeezing Masking (LSM), the Rotating S-boxes Masking (RSM), the Inner Product Masking (IPM), and the most recent Direct Sum Masking (DSM) schemes. After that, we introduce an extension of the DSM to a so-called Multi-shares Direct Sum Masking (DSM).

In Part II, we introduce, in the SCAs' topic, a new concept called spectral approach through two chapters.

Chapter 5 introduces some definitions of the well-known spectral approach of pseudo-Boolean function computations.

In Chap. 6, we generalize a recent spectral approach [20] to significantly speed up the CPA, without any loss in its effectiveness [23]. This improvement holds also for carrying out the CPA incrementally. We will show that this approach is also straightforwardly extendible against the masking-based protected implementations.

In Part III, we combine two notions. First, we introduce the so-called *coalescence principle* and its advantages. Second, we explain its compatibility with the *spectral approach*. Consequently, thanks to the combination of the *coalescence principle* and the *spectral approach*, we introduce several algorithmic improvements in the implementations of CPA, covariance-based distinguisher, LRA, and template attacks. This part is divided into four chapters.

In Chap. 7, the *coalescence principle* is defined and its advantages are explained. Then, especially we prove the asymptotic equivalence between SCAs with and without respect of this principle.

In Chap. 8, thanks to the duo *coalescence principle and spectral approach*, we introduce several significant improvements in the implementation of the LRA [24]. All these improvements are compatible with the incremental implementation of LRA. Furthermore, we extend these improvements also against the masked-based protected implementations. In the sequel, we exhibit for the first time a relation between LRA and the Numerical Normal Form (NNF) [25] of the leakage function. This relation leads to exhibit, also for the first time, the optimal expression of the normalized product combination against the arithmetic masking. All these mathematical results will be validated experimentally by simulations.

In Chap. 9, we follow the same approach (*coalescence principle* then *spectral approach*) for improving the implementation of the Template Attacks (TAs) [26]. First, we carry out a formalization of the TAs without coalescence. Then, we exhibit the formalization of the TAs with coalescence. This formalization leads to significant improvements of the TA implementations thanks to the combination of the Law of Large Numbers (LLN) and of the spectral approach. In the sequel, we introduce a new leakage combining method to make the masked leakage exploitable by the TAs, in general, and especially by our TA version. All these results will be validated experimentally both by synthetic traces (obtained by simulation) and by real-world traces (obtained by actual measurements).

At the end of this book, through Chap. 10, we explain how, following the spectral approach, one can very significantly speed up processing the high-order template attack against any masking scheme.

Finally, we conclude that the coalescence principle leads to significant complexity reductions (time and memory) in several SCAs. These improvements happen, asymptotically, without any loss of effectiveness.

Furthermore, the coalescence principle turns the physical leakage into a pseudo-Boolean function (a real-value function defined over the leakage values, living in $\mathbb{F}_2^n$). Thereby, several more mathematical tools could be used, among them the Walsh-Hadamard Transform (WHT) and the NNF. These tools led to further improvements as will be shown in the rest of the book.

References

1. Kocher PC (1996) Timing attacks on implementations of Diffie-Hellman, RSA, DSS, and other systems. In: Koblitz N (ed) Advances in cryptology - CRYPTO'96, 16th annual international cryptology conference, Santa Barbara, California, USA, August 18–22, 1996, Proceedings. Lecture notes in computer science, vol 1109. Springer, pp 104–113
2. Kocher P, Jaffe J, Jun B (1999) Differential power analysis. In: Advances in cryptology - CRYPTO'99. Springer, pp 388–397
3. Richter B, Wild A, Moradi A (2019) Automated probe repositioning for on-die EM measurements. In: Pan DZ (ed) Proceedings of the international conference on computer-aided design, ICCAD 2019, Westminster, CO, USA, November 4–7, 2019. ACM, pp 1–6
4. Genkin D, Shamir A, Tromer E (2014) RSA key extraction via low-bandwidth acoustic cryptanalysis. In: Garay JA, Gennaro R (eds) Advances in cryptology - CRYPTO 2014 - 34th annual cryptology conference, Santa Barbara, CA, USA, August 17–21, 2014, Proceedings, Part I. Lecture notes in computer science, vol 8616. Springer, pp 444–461
5. Genkin D, Pattani M, Schuster R, Tromer E (2018) Synesthesia: detecting screen content via remote acoustic side channels. CoRR arXiv:abs/1809.02629
6. Tramèr F, Boneh D, Paterson KG (2020) Remote side-channel attacks on anonymous transactions. Cryptology ePrint Archive, Report 2020/220. https://eprint.iacr.org/2020/220
7. Genkin D, Pipman I, Tromer E (2015) Get your hands off my laptop: physical side-channel key-extraction attacks on PCs - extended version. J Cryptogr Eng 5(2):95–112
8. Balasch J, Gierlichs B, Reparaz O, Verbauwhede I (2015) DPA, bitslicing and masking at 1 GHz. In: Güneysu T, Handschuh H (eds) Cryptographic hardware and embedded systems - CHES 2015 - 17th international workshop, Saint-Malo, France, September 13–16, 2015, Proceedings. Lecture notes in computer science, vol 9293. Springer, pp 599–619
9. Genkin D, Pachmanov L, Pipman I, Tromer E (2015) Stealing keys from PCs using a radio: cheap electromagnetic attacks on windowed exponentiation, pp 207–228, 09 2015

10. Mangard S, Oswald E, Popp T (2006) Power analysis attacks: revealing the secrets of smart cards. http://www.springer.com/. Springer, December 2006. ISBN 0-387-30857-1, http://www.dpabook.org/
11. Brier É, Clavier C, Olivier F (2004) Correlation power analysis with a leakage model. In: Joye M, Quisquater J-J (eds) Cryptographic hardware and embedded systems - CHES 2004: 6th international workshop Cambridge, MA, USA, August 11–13, 2004. Proceedings. Lecture notes in computer science, vol 3156. Springer, pp 16–29
12. Heuser A, Rioul O, Guilley S (2014) Good is not good enough - deriving optimal distinguishers from communication theory. In: Batina L, Robshaw M (eds) Cryptographic hardware and embedded systems - CHES 2014 - 16th international workshop, Busan, South Korea, September 23–26, 2014. Proceedings. Lecture notes in computer science, vol 8731. Springer, pp 55–74
13. Carbone M, Tiran S, Ordas S, Agoyan M, Teglia Y, Ducharme GR, Maurine P (2014) On adaptive bandwidth selection for efficient MIA. In: Prouff E (ed) Constructive side-channel analysis and secure design - 5th international workshop, COSADE 2014, Paris, France, April 13–15, 2014. Revised selected papers. Lecture notes in computer science, vol 8622. Springer, pp 82–97
14. Schindler W (2005) On the optimization of side-channel attacks by advanced stochastic methods. In: Vaudenay S (ed) Public key cryptography - PKC 2005, 8th international workshop on theory and practice in public key cryptography, Les Diablerets, Switzerland, January 23–26, 2005, Proceedings. Lecture notes in computer science, vol 3386. Springer, pp 85–103
15. Chari S, Rao JR, Rohatgi P (2002) Template attacks. In: Kaliski BS, Jr., Koç ÇK, Paar C (eds) Cryptographic hardware and embedded systems - CHES 2002, 4th international workshop, Redwood Shores, CA, USA, August 13–15, 2002, Revised papers. Lecture notes in computer science, vol 2523. Springer, pp 13–28
16. TELECOM ParisTech SEN research group. DPA Contest. http://www.DPAcontest.org/
17. Werner S, Kerstin L, Paar C (2005) A model stochastic, for differential side channel cryptanalysis. In: LNCS, vol 3659, CHES (Sept 2005), Edinburgh, Scotland, UK. Springer, pp. 30–46
18. Dabosville G, Doget J, Prouff E (2013) A new second-order side channel attack based on linear regression. IEEE Trans Comput 62(8):1629–1640
19. Lomné V, Prouff E, Roche T (2013) Behind the scene of side channel attacks. In: Sako K, Sarkar P (eds) ASIACRYPT (1). Lecture notes in computer science, vol 8269. Springer, pp 506–525
20. Guillot P, Millérioux G, Dravie B, El Mrabet N (2017) Spectral approach for correlation power analysis. In: El Hajji S, Nitaj A, Souidi EM (eds) Codes, cryptology and information security - 2nd international conference, C2SI 2017, Rabat, Morocco, April 10–12, 2017, Proceedings - in honor of Claude Carlet. Lecture notes in computer science, vol 10194. Springer, pp 238–253
21. NIST/ITL/CSD. Advanced encryption standard (AES). FIPS PUB 197, Nov 2001. http://nvlpubs.nist.gov/nistpubs/FIPS/NIST.FIPS.197.pdf (also ISO/IEC 18033-3:2010)
22. ISO/IEC 18033-3:2010. Information technology – security techniques – encryption algorithms – Part 3: Block ciphers
23. Ouladj M, Guillot P, Mokrane F (2020) Generalized spectral approach to speed up the correlation power analysis. In: Yet another conference on CRYPTography and embedded devices YACCRYPTED'2020, Porquerolles (IGESA Center), France, June 15–19th 2020
24. Ouladj M, Guilley S, Prouff E (2020) On the implementation efficiency of linear regression-based side-channel attacks. In: Constructive side-channel analysis and secure design - 11th international workshop, COSADE 2020, Lugano, Switzerland, October 5–7, 2020, Proceedings
25. Carlet C, Guillot P (1999) A new representation of Boolean functions. In: Fossorier MPC, Imai H, Lin S, Poli A (eds) AAECC. Lecture notes in computer science, vol 1719. Springer, pp 94–103
26. Ouladj M, El Mrabet N, Guilley S, Guillot P, Millérioux G (2020) On the power of template attacks in highly multivariate context. J Cryptogr Eng-JCEN

Part I
Classical Side-Channel Attacks

This part introduces an historical background of side-channel analysis, through three chapters. In Chap. 2, a general background of SCA is provided. In Chap. 3, several distinguishers (attacks' tools) are introduced and explained. Finally, in Chap. 4, the different countermeasures against these attacks are studied.

Foundations of Side-Channel Attacks

2

Let us first adopt some useful notations that will hold for the remainder of the book.

2.1 Notations

In this book, we use capital letters for random variables. A realization of a random variable (*e.g.*, X) is denoted by the corresponding lowercase letter (*e.g.*, x). A *sample* of several observations of X is denoted by $(x_i)_i$. Sometimes it is viewed as a vector. The notation $(x_i)_i \hookleftarrow X$ means the initialization of the set of observations $(x_i)_i$ from X. $\mathbb{E}(X)$ and σ_X denote, respectively, the mean and the standard deviation of X.

The notation $\vec{X}$ denotes column vectors and $\vec{X}[u]$ denotes its uth coordinate. Calligraphic letters will be used to denote matrices, such that the elements of a matrix $\mathcal{M}$ will be denoted by $\mathcal{M}[i][j]$. Furthermore, its uth column is denoted by $\vec{\mathcal{M}}[u]$ and its ith line denoted by $\vec{\mathcal{M}}^{\mathsf{T}}[i]$. The symbols $\cdot^2$ and $\sqrt{\cdot}$, applied to vectors and matrices, denote, respectively, the square and the square root of all the vector/matrix coordinates.

In the following side-channel attacks, we will consider that the adversary targets a single sensitive variable Z. The results can be directly extended to the general case, where several variables are targeted in parallel. The sensitive variable Z depends on a public variable X (typically a plaintext or ciphertext) and on a secret subkey k^* such that $Z = F(X, k^*)$ where $F : \mathbb{F}_2^n \times \mathbb{F}_2^n \longrightarrow \mathbb{F}_2^m$. The bit lengths n and m depend on the cryptographic algorithm and the device architecture. Typically $Z = sbox(X \oplus k^*)$, where $sbox$ denotes a substitution box and $\oplus$ denotes the bitwise addition.

An attack is carried out with N side-channel traces $l_0, \ldots, l_{N-1}$. Each $l_i \hookleftarrow L$ corresponds to the processing of $z_i = F(x_i, k^*)$, such that $x_i \hookleftarrow X$ and $z_i \hookleftarrow Z$. The number of samples per traces (time leakage points) is denoted by D.

© The Author(s), under exclusive license to Springer Nature Switzerland AG 2021
M. Ouladj and S. Guilley, *Side-Channel Analysis of Embedded Systems*,
https://doi.org/10.1007/978-3-030-77222-2_2

2.2 General Framework for Side-Channel Attacks

In order to understand the side-channel attacks, one can analyze Fig. 2.1 introduced in [1]. We explain it as follows. A target device is a physical implementation of a cryptographic algorithm, for example, a smart card, an FPGA, or a computer running the AES block cipher. A side channel is an unintended leakage of some information from a cryptographic device through a physical media [1]. In practice, the power consumption and the electromagnetic radiation are the most exploited side channels. But there are various other ones such as the running time variations, the temperature variations, and the acoustic (sound) emanations. The leaked information from side channels is a *physical observable* that can be measured using, for example, an oscilloscope. Hence, the *physical leakage* is the measured *physical observable* leakage. Let us first assume that during the attack, the device involves a constant secret $k^* \hookleftarrow K$. A side-channel *adversary* can query the device N times (by sending N messages $(x_q)_{1 \leq q \leq N} \hookleftarrow X$) to process some cryptographic operation $(Z_q = F(x_q, k^*))_{1 \leq q \leq N}$ and subsequently getting N physical leakages $(L_q)_{1 \leq q \leq N} \hookleftarrow L$ as traces of D time samples.

Using these N leakage traces, the adversary attempts to recover the secret of the crypto-system such as its key. In practice, for each possible value k of the key, the adversary assumes that the *physical leakage* is, up to an additive noise $(n_q)_{1 \leq q \leq N}$, a deterministic function M defined with the pair (x_q, k). Formally:

$$L_q = M(x_q, k) + n_q. \tag{2.1}$$

Finally, the SCAs are usually carried out in two phases. The fist one named the *preparation phase* consists in profiling and characterizing the leakage using a training device. It is used, for example, in template attack [2], in the profiled side channel based in the LRA [3], and some machine learning-based SCAs [4]. This phase is optional for adversaries. In fact, one can assume that the deterministic part of the leakage follows some model function M as will be detailed in Sect. 2.3. Thereby adversaries could overcome this phase as in the CPA Sect. 3.2.3 and MIA Sect. 3.2.7. In such case, the attack is called non-profiled SCA. As far as evaluators are concerned, this phase is mandatory to properly profile the leakage.

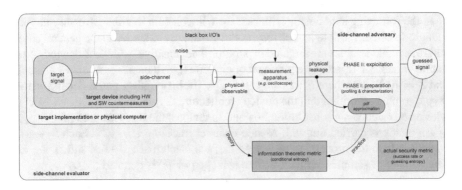

Fig. 2.1 Description of a side-channel key recovery attack [1]

The second phase named the *exploitation phase* consists in guessing the most likely key, using usually one of the statistical methods called distinguishers (see Chap. 3 for more details). The distinguisher requires optionally the results of the preparation phase.

Adversaries need some metrics to compare different side-channel attacks. In practice, they use the success rate and the guessing entropy. Readers can refer to Sect. 2.4 for more details.

In contrast, a side-channel *evaluator* aims at profiling the physical leakage, independently from the subsequent attack kind. In practice, evaluators make use of the information theoretic metric (using the conditional entropy) to compare different implementations [1].

2.3   Leakage Models

Some distinguishers require an exact (or at least an estimated approximate) leakage model. According to [5], the selection (or construction) of a proper leakage model in SCA is at least as important as the selection of a good distinguisher. In practice, if the device responses to the qth query by computing a sensitive value $z_q = F(x_q, k)$, then the leakage function depends only to this value z_q, up to an independent additive noise. Since the first paper about SCA, several generalizations of the leakage model have been introduced [5–8].

2.3.1   Hamming Weight Leakage Model

The Hamming Weight (HW) leakage model has first been suggested by [9, 10]. It assumes that the leakage is proportional to the number of bits set in the sensitive value Z. Formally, there exists some constant real values a and c, such that the qth query corresponds to the sensitive value $z = F(x_q, k)$ and the actual device leakage L_q is related by

$$L_q = a \cdot HW(z) + c + n_q,$$

such that n_q denotes the actual noise.

The Hamming weight model is useful if a pre-charged bus is used [11], e.g., if random access memories (RAM) are addressed. In practice, it is particularly well applicable to software implementations.

2.3.2   Hamming Distance Leakage Model

The Hamming distance leakage model has first been accounted for by the authors of [12]. They state that the dynamic dissipation is not primarily linked to the number of bits set in the sensitive variable Z, but to the number of toggled bits (from 0 to 1 or vice versa). Formally, there exists a constant initial state $z0$, such that

$$L_q = a \cdot HW(z \oplus z0) + c + n_q.$$

Indeed, Hamming weight leakage model is a particular case of Hamming distance model (where $z0 = 0$). In practice, Hamming distance leakage model is particularly applicable to hardware implementations.

It is noteworthy that the Hamming weight/distance leakage models are well suited for most devices, especially for leakage caused by the device registers and data buses.

2.3.3 The Unevenly Weighted Sum of the Bits (UWSB) Leakage Model

In both Hamming weight and Hamming distance models, the leakage is assumed to be equally dependent in each bit of the sensitive variable Z. But, in practice, this perfect resemblance between register/bus bits is unlikely to happen. The Unevenly Weighted Sum of the Bits (UWSB) leakage model is more realistic [13]. Namely, it is an n-variables degree-one polynomial into the Z's bits with real coefficients:

$$L_q = \sum_{i=1}^{n} a_i z_i + c + n_q.$$

Notice that UWSB is a generalization of the Hamming weight/distance leakage model.

2.3.4 Polynomial Leakage Model

The use of the UWSB model is based on the assumption that the leakages related to the manipulation of two different bits are independent (the so-called Independent Bit leakage (IBL) hypothesis). A natural extension of the UWSB model is to consider the leakage model as an n-variables polynomial of the Z's bits of any degree. In fact, the degree of polynomials defined over the Z's bits is at most n. It stems from the fact that any real function defined over n bits (e.g., $\mathbb{F}_2^n$) can be represented by an n-variables polynomial of degree at most equals n (since for any strictly positive integer m one has $z_i^m = z_i$) [5]. Hence, the leakage model:

$$L_q = \sum_{i=1}^{n} a_i z_i + \sum_{1 \leq i < j \leq n} a_{i,j} z_i z_j + \cdots + a_{1,\ldots,n} z_1 \ldots z_n + c + n_q.$$

This is the most general form of the leakage model and it must be considered by designers, in order to study the device resistance against SCA in the worst-case scenario.

2.3.5 Leakage Model in Profiled SCA

In profiled SCA, the adversary is assumed to have an exact replica of the target device. In such a case, the adversary can fit exactly the leakage model. Namely, she can fit different parameters of the leakage probability law (e.g., different moments: mean, variance, covariance, etc.). Examples of profiled SCA are the Template Attacks (TA) and some machine learning-based attacks.

2.4 SCA Security Metrics

To assess a side-channel attack, the adversary should use a security metric. Similarly, the device designers assess the security of an implementation by using a security metric. Several metrics are used in the literature (and are even standardized, as in ISO/IEC 17825 [14]), and are reviewed below.

2.4.1 Success Rate (SR)

One of the most used evaluation metrics in SCA context is the Success Rate (SR). It is defined with an arbitrary order o. The SR of order $o = 1$ (resp. $o = 2, \dots$) relates to the probability that the correct key is the first guessed (resp. one of two first guessed key, $\dots$) by the analyzer handling a given distinguisher [15]. Formally, a success function S_{k*}^{o} of order o, while the correct key equals k^*, is defined as $S_{k*}^{o} = 1$ if k^* belongs to the first o guessed keys after an SCA, $S_{k*}^{o} = 0$ elsewhere. SR is the expectation of the random variable $S_{k*}^{o} = 1$.

2.4.2 Guessing Entropy (GE)

The second evaluation metric is the Guessing Entropy (GE). It is the averaged position of the true key under a given side-channel attack. That is why it is also called the mean rank metric in the literature. Equivalently, it is the average number of guessed keys to check after an SCA has been carrying out [15]. Formally, if i denotes the location of the true key k^* due to an SCA, GE is the expectation of this location i.

2.4.3 Signal-to-Noise Ratio (SNR)

The traces can be characterized according to their *Signal-to-Noise Ratio* (SNR) as defined in [16, Sect. 4.3.2, page 73]). The *monovariate* SNR is defined at each time sample of the trace. Formally, it is defined as

Definition 1 *(Monovariate SNR trace)* Let L be the random variable that represents the leakage, and X be that corresponds to the messages. The SNR is defined by

$$SNR = \frac{\text{VAR}(\mathbb{E}(L|X))}{\mathbb{E}(\text{VAR}(L|X))},$$

(2.2)

where $\mathbb{E}$ denotes the expectation and VAR the variance operators.

We notice that this notion of SNR is useful to predict the approximated number (denoted by $N_{80\%}$) of the required traces to reach 80% of the success rate. Indeed, $N_{80\%}$ is proportional to the inverse of SNR, as shown in [17].

2.4.4 Normalized Inter-Class Variance (NICV)

In some cases, such as in software implementations, the monovariate leakage can feature a strong SNR. However, this gives no intuition regarding the proportion of signal which is informative within the trace. Typically, could the SNR be improved more by reducing further the noise? The Normalized Inter-Class Variance (NICV [18]) allows to answer this question. The NICV is the proportion of the traces variance and which can be explained by

Definition 2 *(Monovariate NICV trace)* Let L and X be as above. The NICV is defined by

$$NICV = \frac{\text{VAR}(\mathbb{E}(L|X))}{\text{VAR}(L)}.$$

(2.3)

Owing to the law of total variance, that is,

$$\text{VAR}(L) = \text{VAR}(\mathbb{E}(L|X)) + \mathbb{E}(\text{VAR}(L|X)),$$

it can be seen that NICV is bounded between 0 and 1, and that $NICV = 1/(1 + 1/SNR)$. This implies that

$$\frac{\partial\ NICV}{\partial\ SNR} = \frac{1}{(1 + SNR)^2} > 0,$$

i.e., NICV is an increasing function of SNR. The attack success rate is thus improved by increasing either the SNR or the NICV.

2.4.5 Information Theory Metric

The disadvantage of success rate and guessing entropy metrics is the fact that they depend on the used SCA method. In order to soundly evaluate any given implementation independently from the adversary's algorithm, one can leverage the mutual information metric [1]. This theoretical metric depends on the leakage probability distribution. Namely, it is based on the conditional probability distribution $P[k|L]$

for computing the mutual information $MI(K, L)$ between the key random variable K and the leakage random variable L.

In practice, there are two caveats in considering this conditional probability. First of all, the conditional probability distribution is not known exactly but it is just approximated through physical observations (so-called estimations or template profiles). Second, the leakage dimension should be reduced, for computational capacity reasons. It is noticeable that SCAs that exploit an approximated probability distribution of a reduced dimension leakage are nothing but a generic template attack [1]. The interested reader is referred to Sect. 3.2.10 for more details.

2.4.6 Metrics in Machine Learning

The most used metrics in the conventional machine learning are *accuracy*, *recall*, and *loss*. *Accuracy* indicates the portion of samples correctly classified; it is used in machine learning-based SCA. *Recall* returns the ratio between true positives and the sum of true positives and false negatives [8]. *Loss* (also called error) is used in Deep Learning (neural network). It indicates the average classification error of the network. Another metrics are used in Deep learning as alternatives such as *sensitivity* and *specificity* [19–21]. It is noteworthy that in [22], it is experimentally shown that conventional deep learning metrics are not very consistent in the context of side-channel analysis [8].

2.4.7 Relation Between the Security Metrics

In addition to the relation between SNR and NICV exhibited in Sect. 2.4.4, recently, authors of [23] establish the link between the probability of success (SR), the guessing entropy, and the mutual information $MI(K, L)$ metrics. Moreover, they provide the relation between these metrics, the signal-to-noise ratio (SNR) of the leakage, and the minimum number of the required queries to reveal the secret subkey.

An important result of this paper is the assertion that in the case of additive Gaussian noise (which is the most common noise type), the use of the SNR is sufficient enough to accurately predict the security of a device.

2.5 Pre-processing of the Leakage Traces for SCA

Since the first papers of SCA, several techniques of pre-processing measurements have been suggested. There goal is to reduce the computation complexity and/or the number of the required measurements for achieving a successful SCA. One can spread these techniques in four parts, discussed hereafter.

2.5.1 Traces Synchronization

When the adversary retrieves the physical traces, they are misaligned either because of an inaccurate triggering of measurements or because of designer countermeasures. Synchronization techniques aim at cancelling this measurement misalignment. As examples of synchronization tools we can cite: (1) the alignment technique suggested in [16] which consists in performing a cross-correlation with sliding window to search a pattern, (2) the elastic alignment technique introduced in [24] using the Dynamic Time Warping (DTW) algorithm and (3) pattern detection technique suggested by [25] which is based on wavelet analysis.

2.5.2 Noise Filtering

Several tools have been suggested in the literature for filtering the noise. One can simply average the traces for filtering. Elsewhere, one can use (1) the fourth-order cumulant as suggested in [26], (2) linear filters as Wiener filter or Kalman filter [27], (3) the Singular Spectrum Analysis (SSA) as a filtering technique [28], (4) denoising side-channel traces based on wavelets introduced in [29], or (5) denoising technique suggested by authors of [25] based on a combination between mutual information and wavelet analysis. Recently, the Independent Components Analysis (ICA) has also been suggested to denoise side-channel traces [30,31].

2.5.3 Points-of-Interest (PoI) Selection

Points of Interest (POI) is the time samples that the adversary could consider to turn the information into a secret recovery. The fist step in most SCAs is the selection of the POI in the leakage measurements [32]. This selection leads to a decrease of the attack complexity, while hopefully also an increase in its success rate.

Since the introduction of the template attacks [2,33], the methods to find a good set of POI have been discussed. The former articles use DPA attacks to find the relevant POI. Subsequently, several PoI detection methods are suggested in the literature. For example, one can take as POI the peaks of a successful attack like DPA, CPA, LRA, or mutual information [34]. Otherwise, we can use techniques based on the T-test [35,36] like the Sum Of Squared pairwise Differences (SOSD) and the Sum of Squared Pairwise T-differences (SOST) [37,38]. One can also use the ANalysis Of VAriance (ANOVA) F-test, which is called Normalized Inter-Class Variance (NICV) in [17,39,40]. Else, one can use the technique based on the computation of the SNR [16].

In a recent work [32], the authors conclude that the correlation-based test (which is based on CPA) is better suited to the detection of POI. This work improves a PoI selection method for higher order DPA [41] using the central (higher order) statistical moment [42].

Nevertheless, limiting a PoI detection to univariate leakages ignores the significant gains that can be obtained with dimensionality reduction techniques (aka projections) [32]. That is precisely what we are studying in the next subsection.

2.5.4   Dimensionality Reduction

In the literature, several suggested methods aim at the best dimensionality reduction. They are automated methods to more efficiently select the POI. The dimensionality reduction can be carried out by the Principal Component Analysis (PCA) [43], the Linear Discriminant Analysis (LDA) [44], or the Projection Pursuit (PP) [30,45].

According to [46], the optimal dimensionality reduction strategy coincides asymptotically with the LDA. It is noticeable that a variant of LDA exists and is called Quadratic Discriminant Analysis (QDA). In QDA, an individual covariance matrix is estimated for every class of observations. The QDA technique is particularly useful when there is prior knowledge that individual classes exhibit distinct covariances. A method compromising between LDA and QDA is the Regularized Discriminant Analysis (RDA). Both QDA and RDA have been studied in the context of machine learning [47], but they have never been applied as an SCA topic.

References

1. Standaert F-X, Malkin T, Yung M (2009) A unified framework for the analysis of side-channel key recovery attacks. In: EUROCRYPT, April 26–30 2009, Cologne, Germany. LNCS, vol 5479. Springer, pp 443–461
2. Chari S, Rao JR, Rohatgi P (2002) Template attacks. In: Kaliski BS, Jr., Koç ÇK, Paar C (eds) Cryptographic hardware and embedded systems - CHES 2002, 4th international workshop, Redwood Shores, CA, USA, August 13–15, 2002, Revised papers. Lecture notes in computer science, vol 2523. Springer, pp 13–28
3. Schindler W (2005) On the optimization of side-channel attacks by advanced stochastic methods. In: Vaudenay S (ed) Public key cryptography - PKC 2005, 8th international workshop on theory and practice in public key cryptography, Les Diablerets, Switzerland, January 23–26, 2005, Proceedings. Lecture notes in computer science, vol 3386. Springer, pp 85–103
4. Masure L, Dumas C, Prouff E (2019) Gradient visualization for general characterization in profiling attacks. In: Constructive side-channel analysis and secure design - 10th international workshop, COSADE 2019, Darmstadt, Germany, April 3–5, 2019, Proceedings, pp 145–167
5. Doget J, Prouff E, Rivain M, Standaert F-X (2011) Univariate side channel attacks and leakage modeling. J Cryptogr Eng 1(2):123–144
6. Duc A, Dziembowski S, Faust S (2014) Unifying leakage models: from probing attacks to noisy leakage. IACR Cryptol ePrint Arch 2014:79
7. Prest T, Goudarzi D, Martinelli A, Passelègue A (2019) Unifying leakage models on a rényi day. In: Advances in cryptology - CRYPTO 2019 - 39th annual international cryptology conference, Santa Barbara, CA, USA, August 18–22, 2019, Proceedings, Part I, pp 683–712
8. Perin G (2019) Deep learning model generalization in side-channel analysis. Cryptology ePrint Archive, Report 2019/978. https://eprint.iacr.org/2019/978

9. Kocher P, Jaffe J, Jun B (1999) Differential power analysis. In: Advances in cryptology - CRYPTO'99. Springer, pp 388–397
10. Messerges TS, Dabbish EA, Sloan RH (1999) Investigations of power analysis attacks on smartcards. In: USENIX — Smartcard'99, May 10–11 1999, Chicago, Illinois, USA, pp 151–162 (Online PDF)
11. Kerstin L, Kai S, Paar C (2004) DPA on n-bit sized Boolean and arithmetic operations and its application to IDEA, RC6, and the HMAC-construction. In: CHES, August 11–13, Cambridge, MA, USA. Lecture notes in computer science, vol 3156. Springer, pp. 205–219
12. Brier É, Clavier C, Olivier F (2004) Correlation power analysis with a leakage model. In: Joye M, Quisquater J-J (eds) Cryptographic hardware and embedded systems - CHES 2004: 6th international workshop Cambridge, MA, USA, August 11–13, 2004. Proceedings. Lecture notes in computer science, vol 3156. Springer, pp 16–29
13. Zheng Y, Zhou Y, Yu Z, Hu C, Zhang H (2014) How to compare selections of points of interest for side-channel distinguishers in practice? In: Hui LCK, Qing SH, Shi E, Yiu S-M (eds) Information and communications security - 16th international conference, ICICS 2014, Hong Kong, China, December 16-17, 2014, Revised selected papers. Lecture notes in computer science, vol 8958. Springer, pp 200–214
14. ISO/IEC JTC 1/SC 27/WG 3. ISO/IEC 17825:2016: information technology – security techniques – testing methods for the mitigation of non-invasive attack classes against cryptographic modules. https://www.iso.org/standard/60612.html
15. Standaert F-X, Gierlichs B, Verbauwhede I (2008) Partition vs. comparison side-channel distinguishers: an empirical evaluation of statistical tests for univariate side-channel attacks against two unprotected CMOS devices. In: ICISC, December 3–5 2008, Seoul, Korea. LNCS, vol 5461. Springer, pp 253–267
16. Mangard S, Oswald E, Popp T (2006) Power analysis attacks: revealing the secrets of smart cards. Springer, Berlin. ISBN 0-387-30857-1, http://www.dpabook.org/
17. Bhasin S, Danger J-L, Guilley S, Najm Z (2014) Side-channel leakage and trace compression using normalized inter-class variance. In: Proceedings of the 3rd workshop on hardware and architectural support for security and privacy, HASP'14, New York, NY, USA. ACM, pp 7:1–7:9
18. Bhasin S, Danger J-L, Guilley S, Najm Z (2014) NICV: normalized inter-class variance for detection of side-channel leakage. In: International symposium on electromagnetic compatibility (EMC'14/Tokyo). IEEE, May 12–16 2014. Session OS09: EM Information Leakage. Hitotsubashi Hall (National Center of Sciences), Chiyoda, Tokyo, Japan
19. Timon B (2019) Non-profiled deep learning-based side-channel attacks with sensitivity analysis. IACR Trans Cryptogr Hardw Embed Syst 2019(2):107–131 Feb
20. Simonyan K, Vedaldi A, Zisserman A (2013) Deep inside convolutional networks: visualising image classification models and saliency maps. CoRR arXiv:abs/1312.6034
21. Shrikumar A, Greenside P, Kundaje A (2017) Learning important features through propagating activation differences. ArXiv arXiv:abs/1704.02685,
22. Picek S, Heuser A, Jovic A, Bhasin S, Regazzoni F (2018) The curse of class imbalance and conflicting metrics with machine learning for side-channel evaluations. IACR Trans Cryptogr Hardw Embed Syst 2019(1):209–237 Nov
23. de Chérisey E, Guilley S, Rioul O, Piantanida P (2019) Best information is most successful mutual information and success rate in side-channel analysis. IACR Trans Cryptogr Hardw Embed Syst 2019(2):49–79
24. van Woudenberg JGJ, Witteman MF, Bakker B (2011) Improving differential power analysis by elastic alignment. In: Kiayias A (ed) CT-RSA. Lecture notes in computer science, vol 6558. Springer, pp 104–119
25. Debande N, Souissi Y, Elaabid MA, Guilley S, Danger J-L (2012) Wavelet transform based preprocessing for side channel analysis. In: 45th annual IEEE/ACM international symposium on microarchitecture, MICRO 2012, workshops proceedings, Vancouver, BC, Canada, December 1–5, 2012. IEEE Computer Society, pp 32–38

26. Le T-H, Cledière J, Servière C, Lacoume J-L (2007) Noise reduction in side channel attack using fourth-order cumulant. IEEE Trans Inf Forensics Secur 2(4):710–720. https://doi.org/10.1109/TIFS.2007.910252 December
27. Souissi Y, Guilley S, Danger J-L, Duc G, Mekki S (2010) Improvement of power analysis attacks using Kalman filter. In: ICASSP, IEEE Signal Processing Society, March 14–19 (2010), Dallas, TX, USA. IEEE, pp. 1778–1781. https://doi.org/10.1109/ICASSP.2010.5495428
28. Del Pozo SM, Standaert F-X (2015) Blind source separation from single measurements using singular spectrum analysis. In: Güneysu T, Handschuh H (eds) Cryptographic hardware and embedded systems - CHES 2015 - 17th international workshop, Saint-Malo, France, September 13–16, 2015, Proceedings. Lecture notes in computer science, vol 9293. Springer, pp 42–59
29. Pelletier H, Charvet X (2005) Improving the DPA attack using wavelet transform, September 26–29 2005. Honolulu, Hawai, USA; NIST's physical security testing workshop. Website: http://csrc.nist.gov/groups/STM/cmvp/documents/fips140-3/physec/papers/physecpaper14.pdf
30. Maghrebi H, Prouff E (2018) On the use of independent component analysis to denoise side-channel measurements. In: Fan J, Gierlichs B (eds) Constructive side-channel analysis and secure design - 9th international workshop, COSADE 2018, Singapore, April 23–24, 2018, Proceedings. Lecture notes in computer science, vol 10815. Springer, pp 61–81
31. Le T-H, Cledière J, Servière C, Lacoume J-L (2007) How can signal processing benefit side channel attacks? In: Proceedings of IEEE workshop on signal processing applications for public security and forensics (SAFE), pp 1–7, April 11–13 2007, Washington D.C., USA
32. Durvaux F, Standaert F-X (2016) From improved leakage detection to the detection of points of interests in leakage traces, pp 240–262, 05 2016
33. Rechberger C, Oswald E (2004) Practical template attacks. In: WISA, August 23–25 2004, Jeju Island, Korea. LNCS, vol 3325. Springer, pp 443–457
34. Ou C, Lam SK, Jiang G (2019) The art of guessing in combined side-channel collision attacks. IACR Cryptol ePrint Arch 2019:690
35. Schneider T, Moradi A (2016) Leakage assessment methodology. J Cryptogr Eng 6:02
36. Ding AA, Chen C, Eisenbarth T (2016) Simpler, faster, and more robust t-test based leakage detection, vol 9689, pp 163–183, 04 2016
37. Gierlichs B, Lemke-Rust K, Paar C (2006) Templates vs. stochastic methods. In: CHES, October 10–13 2006, Yokohama, Japan. LNCS, vol 4249. Springer, pp 15–29
38. Zotkin Y, Olivier F, Bourbao E (2018) Deep learning vs template attacks in front of fundamental targets: experimental study. IACR Cryptol ePrint Arch 2018:1213
39. Choudary O, Kuhn MG (2013) Efficient template attacks. In: Francillon A, Rohatgi P (eds) Smart card research and advanced applications - 12th international conference, CARDIS 2013, Berlin, Germany, November 27–29, 2013. Revised selected papers. LNCS, vol 8419. Springer, pp 253–270
40. Danger J-L, Debande N, Guilley S, Souissi Y (2014) High-order timing attacks. In: Proceedings of the 1st workshop on cryptography and security in computing systems, CS2'14, New York, NY, USA. ACM, pp 7–12
41. Durvaux F, Standaert F-X, Veyrat-Charvillon N, Mairy J-B, Deville Y (2015) Efficient selection of time samples for higher-order DPA with projection pursuits. In: Mangard S, Poschmann AY (eds) Constructive side-channel analysis and secure design - 6th international workshop, COSADE 2015, Berlin, Germany, April 13–14, 2015. Revised selected papers. Lecture notes in computer science, vol 9064. Springer, pp 34–50
42. Prouff E, Rivain M, Bevan R (2009) Statistical analysis of second order differential power analysis. IEEE Trans Comput 58(6):799–811
43. Jolliffe IT (2002) Principal component analysis. Springer series in statistics. ISBN: 0387954422
44. Fisher RA (1936) The use of multiple measurements in taxonomic problems. Ann Eugen 7:179–188, 01 (1936)

45. Friedman J, Tukey JW (1974) A projection pursuit algorithm for exploratory data analysis. IEEE Trans Comput c-23:881–889, 10 (1974)
46. Bruneau N, Guilley S, Heuser A, Marion D, Rioul O (2015) Less is more - dimensionality reduction from a theoretical perspective. In: Güneysu T, Handschuh H (eds) Cryptographic hardware and embedded systems - CHES 2015 - 17th international workshop, Saint-Malo, France, September 13–16, 2015, Proceedings. Lecture notes in computer science, vol 9293. Springer, pp 22–41
47. Wu W, Mallet Y, Walczak B, Penninckx W, Massart DL, Heuerding S, Erni F (1996) Comparison of regularized discriminant analysis linear discriminant analysis and quadratic discriminant analysis applied to NIR data. Anal Chim Acta 329(3):257–265

Side-Channel Distinguishers

3

A distinguisher is a statistical tool whose purpose is to determine the most probable key among a set of keys. Several distinguishers are introduced in the literature. Hereafter, we present the most used ones.

3.1 SCA Distinguishers Classification

Several classifications of distinguishers can be considered. One can split them into two classes, monovariate and multivariate distinguisher, according to the number of the considered (time) samples during the distinguisher computation. Some authors split the distinguishers in several classes according to the type of targeted leakage (time, power consumption, electromagnetic emanation, …). Another way to classify is to separate between distinguishers according to the assumption that the adversary owns a copy of the target device before the attack or not (profiled and non-profiled attacks, respectively). In fact, if the adversary has an equivalent device to the target one, then she can profile the leakage before the attack phase. The profiling phase often consists in estimating the leakage probability distribution, namely, by estimating different statistical moments (Means, (co)variance, …). Whereas the attack phase consist in computing some "distance" between the distribution of the actual leakage and the estimated leakage distribution. Different "distances" are possible such as the square norm of the difference (as in LRA) and notably the maximum of the likelihood. Finally, still other classifications of side-channel attacks are possible as explored in [1].

In this book, the distinguishers are classified into two main classes. The first one contains all the distinguishers that target the non-protected implementations, especially non-masked ones. It is called the class of first-order distinguishers. The second class contains the attacks that target the protected implementations by masking. It is called the class of higher order distinguishers.

© The Author(s), under exclusive license to Springer Nature Switzerland AG 2021
M. Ouladj and S. Guilley, *Side-Channel Analysis of Embedded Systems*,
https://doi.org/10.1007/978-3-030-77222-2_3

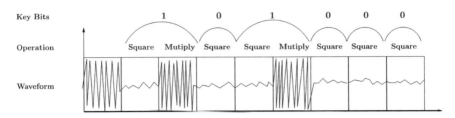

Fig. 3.1 SPA against the RSA exponentiation

3.2 First-Order Distinguishers

Several first-order distinguishers are introduced in literature. Hereafter, the most known distinguishers are described.

3.2.1 Simple Power Analysis (SPA)

Simple power analysis is a straightforward analysis of the leakage trace during a sensitive computation. A classical example of such attacks is the key revealing of RSA protocol by analyzing a temporal trace of a device power consumption, during the RSA exponentiation [2]. In practice, the efficient computation of the exponentiation of the message by the key k^* is carried out by the square and multiply algorithm. Since the consumption of the device during the squaring operations and multiplications is different, one can reveal the key bit by bit as shown in Fig. 3.1.

3.2.2 Differential Power Analysis (DPA)

Differential power analysis is an SCA introduced by Kocher et al. in [3]. It consists in analyzing the relation between the consumption and the manipulated data, using the so-called difference of means. Namely, the adversary splits the set of leakage in two subsets P_0 and P_1, according to the first bit (for example) of the guessed sensitive value. In fact, the guessed sensitive value depends on the guessed subkey. A good guess of the subkey leads to a good partition of the traces. Only the good partition of the traces (good guess of the subkey) leads to a big difference between the mean of P_0 and that of P_1.

This paper is the real beginning of the side-channel topic. Even in practice this attack gives several ghost peaks (false positive), it prompted researchers to introduce several more efficient attacks. Some generalization of this approach has been introduced in literature such as the Multi-bit DPA [4] and the Partitioning Power Analysis [5,6].

3.2.3 Correlation Power Analysis (CPA)

If the adversary owns the leakage model denoted by $\mathbf{M}$, she can lead an efficient SCA by computing the Pearson correlation coefficient [7] between the physical leakage $\mathbf{L}$ and the leakage model $\mathbf{M}$ applied to the key k:

$$\rho_{\mathbf{L},\mathbf{M_k}} = \frac{cov(\mathbf{L}, \mathbf{M_k})}{\sigma_\mathbf{L} . \sigma_{\mathbf{M_k}}} = \frac{\mathbb{E}\left((\mathbf{L} - \overline{\mathbf{L}})(\mathbf{M_k} - \overline{\mathbf{M_k}})\right)}{\sigma_\mathbf{L} . \sigma_{\mathbf{M_k}}}. \tag{3.1}$$

This statistical tool estimates how much two random variables (leakage and model) are linearly dependent. The good guess of the key leads to the highest value of the coefficient. This approach is introduced in the SCA topic by Coron et al. in [8], then used in [9] and called the Correlation Power Analysis (CPA) in [10].

CPA Implementation Efficiency

According to [11], the Pearson coefficient defined in Eq. (3.1) is equivalent to

$$\rho(\mathbf{M_k}, \mathbf{L}) = \frac{N s_5 - s_1 s_3}{\sqrt{(N s_2 - s_1^2)(N s_4 - s_3^2)}}, \tag{3.2}$$

where $s_1 = \sum_{i=1}^{N} L(x_i), s_2 = \sum_{i=1}^{N} L(x_i)^2, s_3 = \sum_{i=1}^{N} M(x_i, k), s_4 = \sum_{i=1}^{N} M(x_i, k)^2,$

and $s_5 = \sum_{i=1}^{N} M(x_i, k) L(x_i).$

The advantage of this second definition of the Pearson correlation coefficient is its possible incremental computation. Namely, if the adversary computes this coefficient for a set of N_1 traces without revealing the secret, she can update the computation by adding another set of N_2 traces without recomputing Eq. (3.2) for the first set.

As stated in [11], if the number of traces equals N and the number of the subkey equals 2^n, then the computations of the sums ($s_1 \ldots s_5$) can be carried out with an overall complexity of $\mathscr{O}(N 2^n)$.

Also recently, a spectral approach is introduced [12] to speed up the CPA computations. Thanks to this approach introduced by Guillot et al., if the adversary uses exactly one trace per possible message value (2^n messages), then the complexity decreases from $\mathscr{O}(2^{2n})$ to $\mathscr{O}(n 2^n)$. The disadvantage of this approach is the specific required set of measurements. In Chap. 6, we provide a generalization of this approach to any set of measurements. Our approach leads to a linear overall complexity $\mathscr{O}(N)$ instead of $\mathscr{O}(N 2^n)$, when $N \gg n 2^n$ (which is not uncommon in practice). Moreover, this improvement is straightforwardly applicable to the covariance-based distinguisher, Sect. 3.2.5.

3.2.4 Rank-Based CPAs

They are based on nonparametric statistics, namely, on the so-called nonparametric correlation coefficients. In contrast to the native CPA, they do not use the exact values of data but only their rank [13]. Examples of nonparametric correlation coefficients are Spearman's ρ, Kendal's τ, and Gini Γ coefficients. Unlike Pearson's correlation coefficient, rank correlation does not assume a linear relationship between the model and physical leakage. So the ordinary CPA based on the Pearson correlation coefficient requires more information in the data than Spearman's, for example, since the last one only assume that the data is ordered [14]. A comparative evaluation of rank-based correlation DPA, on an AES prototype Chip, is carried out in [14]. It is the first article that uses the Spearman rank correlation coefficient in SCA.

3.2.5 Covariance-Based Distinguisher

If the adversary owns the leakage model, she can also lead an efficient SCA by computing the covariance between the physical leakage $\mathbf{L}$ and the leakage model $\mathbf{M}$ applied to the key k [15]:

$$COV_{\mathbf{L},\mathbf{M_k}} = cov(\mathbf{L}, \mathbf{M_k}) = \mathbb{E}\left((\mathbf{L} - \overline{\mathbf{L}})(\mathbf{M_k} - \overline{\mathbf{M_k}})\right). \tag{3.3}$$

The good guess of the key leads to the highest value of the covariance. In fact, this distinguisher is none other than the numerator for the Pearson coefficient. Similar to the CPA, one can compute an incremental covariance distinguisher by

$$COV_{\mathbf{L},\mathbf{M_k}} = \frac{1}{N}\sum_{i=1}^{N}(L(x_i) - \overline{L})(M(x_i, k) - \overline{M}_k) = Ns_5 - s_1 s_3. \tag{3.4}$$

Thereby, our improvements which will be introduced in Chap. 6 hold both for the CPA and the covariance-based distinguisher.

3.2.6 Collision Side-Channel Attacks

Collision means that the same leakage occurs in two different time samples (e.g., when the circuit uses a module more than once as an *Sbox*) [16,17]. Instead of matching a leakage with the model, one can match both leakages together. A simple example is when two *sbox* of AES take out the same value then leak the same signal up to a noise. They take out the same value means $sbox_1(x_1 \oplus k_1^*) = sbox_2(x_2 \oplus k_2^*)$ and this is equivalent to $k_1^* \oplus k_2^* = x_1 \oplus x_2$. So, if the adversary sends several messages to the device that process the AES, such that the sub-messages x_1 and x_2 make $x_1 \oplus x_2$ constant, then she can assess if $k_1^* \oplus k_2^* = x_1 \oplus x_2$ using a distinguisher (like CPA) between both leakages.

Other collision-based side-channel attacks have been conducted against public-key crypto-systems. As example, reader can see an attack on modular exponentiation as in [18] and on elliptic curves as in [19].

3.2.7 Mutual Information Analysis (MIA)

It is a (generic) side-channel distinguisher introduced by Gierlichs et al. in [20]. Recall that the CPA assesses the possible linear relation between the adversary's model and the physical leakage. In contrast, MIA is a generic distinguisher in terms that it can capture any type of dependency between the models and the physical leakage [21]. Formally, the CPA's adversary successes if $\hat{\rho}(\mathbf{M_{k^*}}, \mathbf{L}) > \hat{\rho}(\mathbf{M_k}, \mathbf{L})$ for every $k \neq k^*$, while a successful MIA requires only that $\hat{I}(\mathbf{M_{k^*}}, \mathbf{L}) > \hat{I}(\mathbf{M_k}, \mathbf{L})$ for every $k \neq k^*$. This means in particular cases MIA is more efficient than the CPA.

In practice, to carry out an MIA, the adversary has (1) to estimate the leakage Probability Density Functions (PDF) for each possible key-dependent models and (2) to test the dependence of these models with the physical measurements [22]. Still, such model can be inaccurate in practical situations, in particular, because the model must be non-injective for MIA to be sound. Therefore, MIA is not necessarily the most efficient distinguisher [23]. Elsewhere, if the used discrete models correspond perfectly to the actual leakages, then the MIA is a good tool [22,24].

The extension of MIA to the multivariate context (against masking) has been carried out [25,26]. In addition, several in-depth studies on MAIs have been carried out, as in [27–29] and recently in [30] where the optimality and practicability of MIA have been targeted.

In the sequel, two distinguishers will be introduced as competitors of MAI. Namely, they are the Kolmogorov-Smirnov test [31] and the χ^2−Test [32].

3.2.8 Kolmogorov-Smirnov Distance (KS)-Based Distinguisher

In contrast to MIA, an explicit PDF estimation is not always necessary. There is other statistical tools to compare two distributions directly from their values. As example of such alternatives is the Kolmogorov-Smirnov test (KS) [22]. Both of mutual information and Kolmogorov-Smirnov test evaluate some "distance" between two distributions. But, KS-test consists only to evaluate empirical Cumulative Distributions Functions (CDFs) (that of the physical leakage random variables L and those of the conditional distribution of $L|M_k$ for each key dependent), instead of estimating their PDFs.

Formally, the CDF of a random variable X according to its sample $(x_i)_{1 \leq i \leq N}$ is denoted by F_X and defined for each real value t (for threshold) by $F_X(t) = \frac{1}{N} \sum_{i=1}^{N} \delta_{(x_i \leq t)}$, such that the indicator $\delta_{(x_i \leq t)}$ equals 1 if $x_i \leq t$ and 0 elsewhere. The KS-test measures the distance between two random variables X and Y by $sup_t |F_X(t) - F_Y(t)|$. And the KS-inspired distinguisher is defined by averaging

KS-test over the possible values m of the model as follows:

$$KS(L||L|M_k = m) = \mathbb{E}\left[sup_t|F_L(t) - F_{L|M_k=m}(t)|\right].$$

This distinguisher is expected to reach its maximum if the correct key is guessed.

It can also be extended straightforwardly to the multivariate context (against masking) [31]. Another distinguisher similar to MIA but does not need the PDFs (similar to KS-test) is the Cramer-von Mises Criterion [34].

3.2.9 Chi-Squared Test (Chi-2-Test)-Based Distinguisher

After the implementation, the first analysis that the cryptographic device undergoes by the designer is the leakage detection. Namely it consists in testing the independence of two random variables from statistical samples. The random variables are the actual physical leakage and the corresponded theoretic values deduced from the distribution model. The leakage detection is often conducted by the $t-$test [35,36], but recently another method using χ^2-Test is suggested to conduct this detection [37]. The χ^2-Test is based on the computation of some distance between (1) the frequencies of the joint distribution of both random variables and (2) the product of the frequencies of both distributions. For more details, reader can see [37].

In the same paper [37], the authors also show that χ^2-Test can be used as a distinguisher similarly to MIA. For the current use of χ^2-test and MIA, there is no direct relation. It is noticeable that the χ^2-test might not necessarily be the best attack but it is easy to apply and does not need setting [32]. There are differences in the application of the tests. In fact, statistics provides several other tools for testing the independence of two random variables. As examples, one can include the Student, Hotelling, and even nonparametric statistic tests like the Wald-Wolfowitz test [33].

3.2.10 Template Attack (TA)

All the distinguishers introduced so far are used for non-profiled attacks. In contrast, the Template Attack (TA) [38] is a profiled one. This attack is split into two phases: a profiling and a matching stage. During the profiling stage, the adversary has to estimate the PDF of the leakage using a copy of the target device. Often the adversary can assume that the leakage variable follows a multivariate Gaussian (normal) distribution. So, during the profiling phase one needs only to estimate the mean and the covariance matrix. To recover the key, the adversary applies the maximum likelihood principle during the matching phase. Template Attacks are the most powerful attacks from the information-theoretic point of view [39]. A further explanation and several optimizations of the TA will be provided in Chap. 9.

It is noteworthy that even if the leakage don't follow the Gaussian distribution, the adversary can estimate different needed probabilities using Bayes' theorem (during the profiling phase) [24]. During the matching phase, the maximum likelihood principle is useful regardless of the leakage distribution.

3.2.11 Linear Regression-Based Side-Channel Attacks (LRA)

An important weakness of the TA is the large number of measurements required during the profiling phase. This requirement is due to the need of an accurate estimation of the probability densities of different samples of the leakage [40]. To overcome this inconvenience, Schindler introduced a new stochastic approach that consists in approximating probability densities [41]. In fact, this stochastic approach is less efficient than the Template Attack (TA), but it has the advantage that, in the profiling stage, it requires less traces than TA [40] and does not require the knowledge of the secret key. Another advantage is that LRA can be turned into a non-profiled attack [42]. An explanation and several improvements of the LRA implementation will be provided in Chap. 8.

3.2.12 Machine Learning-Based Distinguishers

As we can deduce from the template attacks, the SCAs can be viewed as a classification problem. So, several machine learning techniques can be used as distinguishers. The side-channel researchers have mainly investigated Support Vector Machines (SVMs) [43], Random Forest [44], but also other approaches are studied. For more details, the reader is referred to the recently published survey of machine learning techniques' application in SCAs [45].

Recently, other deep learning approaches are introduced in the SCA area [46]. For example, the Multi-Layer Perceptron (MLP) and Convolutional Neural Network (CNN). The advantage of the CNN compared to MLP is that the CNN has a translation-invariance property. Therefore, CNN is interesting when dealing with de-synchronized side-channel traces [47]. The deep learning-based SCAs can be a profiled attack as in [46,47] or a non-profiled attack as in [48].

3.3 Higher Order Distinguishers

As will be studied in Sect. 4.2, a dth-order masking consists in splitting each sensitive value to at least $d + 1$ shares. If the dth-order masking is perfect (Definition 1), then for a successful SCA the adversary must consider at least $d + 1$ time samples of the leakages that correspond to the shares. Such attack (called the Higher Order attack [49]) need a higher order distinguisher.

To conduct a higher order attack, the adversary can either (1) use the "first-order" distinguishers like those introduced above, but she needs to combine the $d + 1$ leakages before the computation of the distinguisher, or (2) use the joint distribution of the $d + 1$ leakages without a prerequisite combination.

3.3.1 Higher Order Distinguishers Overs Combination

The "first-order" distinguishers can match a model to a leakage of one sensitive variable. Thereby, one needs to combine the leakages corresponding to both shares into one (to form a new exploitable signal), before using such distinguishers. In order to perform a higher order attack, many combination functions are suggested in the literature, such as product combination [50], absolute difference combination (possibly raised to some power) [51], and sine-based combination [52].

In fact, the application of the combination functions leads to an information loss. But, for high noise this loss vanishes and the higher order CPA with the normalized product combination function [53] becomes (nearly) equivalent to the maximum likelihood distinguisher applied to the joint distribution [54].

According to [53], the optimal combination technique against the first-order Boolean masking is the *normalized* product (a monovariate combined leakage defined by $(L^{(0)} - \mathbb{E}(L^{(0)}))(L^{(1)} - \mathbb{E}(L^{(1)}))$). The authors of [53] show that this combination function should be accompanied by $\mathbb{E}((M_k^{(0)} - \mathbb{E}(M_k^{(0)}))(M_k^{(1)} - \mathbb{E}(M_k^{(1)})) \mid Z)$ as an optimal model. Furthermore, authors of [55] state that the CPA with *normalized* product combination is optimal for high noise, independently from the masking technique and especially independently from the masking order. While the attack is carried out on the combination leakage, as suggested in [53,55], all our improvements that will be introduced can straightforwardly be applied (both for CPA/Covariance, LRA, and even for TA).

3.3.2 Higher Order Distinguishers Without a Prerequisite Combination

Authors of [54] showed experimentally that combination functions lose information, and only for high noise, the higher order CPA with the *normalized* product tends toward the maximum likelihood distinguisher applied to the joint distribution [56]. Furthermore, the product of two or more Gaussian distributions is not a Gaussian one, while CPA is only suitable for Gaussian's. The target Higher Order Optimal Distinguisher (HOOD) is introduced in [56] and studied for low and high noise. It is a maximum likelihood-based distinguisher, but does not require a prerequisite combination.

In this section, we introduce the notations. Let us assume an attacker knows the plaintext t and guesses a constant secret key k. They are inputted in a cryptographic algorithm, and their combination gives rise to a *sensitive variable z*. For example, $z(t, k) = \text{Sbox}(t \oplus k)$, in the case of the attack of a substitution box (in block ciphers such as AES or PRESENT). In general, the function $k \mapsto z(t, k)$ is an injection, for all t.

The sensitive variable z is the target of side-channel attacks. In masking schemes, each z is randomly splitted. Namely, dth higher order masking consists in computing each sensitive variable $z(t, k)$ separately in $N = d + 1$ shares.

For example, in dth-order Boolean masking scheme, z is splitted in $(z^{(0)}, \dots, z^{(d)})$, such that

$$z = \bigoplus_{w=0}^{d} z^{(w)}. \tag{3.5}$$

In Eq. (3.5), the $d + 1$ shares are traditionally randomly chosen from d random numbers called masks m_w, with $1 \leq w \leq d$, according to $z^{(1)} = m_1, z^{(2)} = m_2, \dots$ $z^{(d)} = m_d, z^{(0)} = z \oplus \bigoplus_{w=1}^{d} z^{(w)}$. Besides, in Eq. (3.5), the operation is the bitwise XOR, denoted as "$\oplus$".

In general, such masking adapts to other situations. Typically, one just needs the combination to be a group operation. In the rest of this section, we shall consider that this group is additive. We denote it as $(\mathbb{S}, +)$. Notice that multiplicative groups also allow to derive multiplicative masking; therefore, the "$+$" sign shall be understood broadly, as the group $\mathbb{S}$ inner operation. Examples of groups $\mathbb{S}$ are $(\mathbb{F}_2^n, \oplus)$ used in Boolean masking, or $(\mathbb{F}_{2^n}, +)$, such that $+$ is the 2^n-modular addition, which is used in arithmetic masking.

Attack on High-Order Boolean Masking

Template attacks have been introduced as multivariate attacks on unprotected devices in [38]. Their extension to masked implementation has been suggested in [52, 57]. However, template attacks on masking scheme where each share is profiled independently is not clearly described in the state of the art. We discuss it in this chapter, which gives the most general attack compared to the existing literature.

During the profiling phase, the adversary can see the leakage measurement X_q as $(d + 1)$ disjoint sub-traces denoted by $X_q^{(w)}, 0 \leq w \leq d$. Each sub-trace $X_q^{(w)}$ is a sub-trace of length $D^{(w)}$ and corresponds to the leakage of share $z^{(w)}$. We assume that for each w, $0 \leq w \leq d$, the adversary estimates the multivariate Probability Density Function (PDF) of $X_q^{(w)}$, such that she profiles the device consumption of $D^{(w)}$ samples for the wth share.

Thanks to the profiling result, the adversary can estimate the probability $p(X_q^{(w)} | z^{(w)})$ for each manipulated data $z^{(w)}$. We recall that $z^{(w)}$, for $1 \leq w \leq d$, is equal to mask m_w, and $z^{(0)}$ is equal to $z \oplus m_1 \oplus \cdots \oplus m_d$.

According to [55, Thm 7], when attacking $Q > 0$ traces, the $(d + 1)$th-order optimal distinguisher is (for simplicity, we assume that all the mask values have the same probability)

$$\mathscr{D}_{opt}^{d} = \underset{k}{\mathrm{argmax}} \prod_{q=1}^{Q} \sum_{m_1, \dots, m_d \in \mathbb{S}} p(X_q^{(1)} | m_1) \dots p(X_q^{(d)} | m_d) p(X_q^{(0)} | z(t_q, k) \oplus \bigoplus_{w=1}^{d} m_w). \tag{3.6}$$

This equation is the maximum likelihood estimator. It takes as inputs

- the known plaintexts $(t_q)_{1 \leq q \leq Q}$ and
- the leakage sub-traces $(X_q^{(0)}, \dots, X_q^{(d)})_{1 \leq q \leq Q}$,

and returns the most likely key $\hat{k}$ according to the relationship between z, t, and k.

In fact, in [55], the HOOD is studied by taking only one sample per share. In our study, we generalize it by assuming $D^{(w)}$ samples for the wth sub-trace, $0 \leq w \leq d$, and considering this sub-trace as a random vector. That means, we generalize the HOOD in same way when one goes from the first-order DPA (one sample per trace to compute the distinguisher) to the (first-order) template attack (D samples per trace but concerning only one sensitive variable z). Here, we migrate from the higher order DPA to the higher order template attack.

To compute our distinguisher efficiency, one can see that

$$\mathcal{D}_{opt}^d = \operatorname*{argmax}_k \prod_{q=1}^Q Sum_q^{BM}(z(t_q, k)) = \operatorname*{argmax}_k \sum_{q=1}^Q \log Sum_q^{BM}(z(t_q, k)), \quad (3.7)$$

where Sum_q^{BM} is a pseudo-Boolean function (meaning a real function defined over Boolean vectors) of 2^n values, equal to

$$Sum_q^{BM}(z) = \sum_{m_1,\ldots,m_d \in \mathbb{S}} p(X_q^{(1)}|m_1)\ldots p(X_q^{(d)}|m_d) \atop p(X_q^{(0)}|z \oplus m_1 \oplus \cdots \oplus m_d) \quad (3.8)$$

$$= \sum_{m_1 \in \mathbb{S}} \cdots \sum_{m_d \in \mathbb{S}} p(X_q^{(1)}|m_1)\ldots p(X_q^{(d)}|m_d) \atop p(X_q^{(0)}|z \oplus m_1 \oplus \cdots \oplus m_d) \quad (3.9)$$

$$= p(X_q^{(1)}|.) \otimes \cdots \otimes p(X_q^{(d)}|.) \otimes p(X_q^{(0)}|.)(z),$$

such that $\otimes$ is the convolution product relatively to the law $\oplus$ in the group $(\mathbb{S}, \oplus)$. Clearly, the property that the value of z can be recovered as $z = z^{(0)} - \sum_{w=1}^d z^{(w)}$ is central to apparition of the convolution. This property is specific to the Boolean masking, but extends to all masking schemes. Indeed, in all masking schemes, there is a relationship between the shares $z^{(0)}, \ldots, z^{(d)}$ which allows to recover the sensitive value $z(t, k)$.

3.4 Comparison of Distinguishers

Several side-channel attacks are asymptotically equivalent, when the adversary targets only one time sample per trace. In particular, in [42], the authors prove that most univariate attacks suggested in the literature can be expressed as a CPA with different leakage models. Indeed, both Differential Power Analysis (DPA), Partitioning Power Analysis (PPA), and CPA attacks can be reformulated to reveal a correlation coefficient computation. They only differ in the involved model [42]. In [58], the authors show that the amount of information leaked by a cryptographic device measured with an information-theoretic metric is connected to the correlation coefficient.

In the other hand, authors of [59] state that in a standard situation (when the leakage follows almost perfectly the Hamming weight model) CPA is more efficient than MIA. In fact, the KSA efficiency is similar to that of MIA [31], but KSA is less robust to the model degradation than MIA [59].

According to a recent work [60], if the leakage probabilities are totally known, the best possible side-channel attack corresponds to the Maximum Likelihood (ML)-based distinguisher. Elsewhere, if the profiling is unavailable, MIA coincides with the ML only if the leakage probabilities are replaced by their online estimations. In fact, MIA outperforms CPA when (1) the leakage model is known but the noise is not Gaussian or (2) the leakage model is partially unknown and if the noise is Gaussian. In the last case (without a good model), the LRA is also preferred than the CPA [42,61].

In brief, if the noise is Gaussian (which is a common assumption), for the non-profiled analysis, CPA is almost optimal when the leakage model is known up to an affine transformation [15]. Elsewhere, the LRA is preferred [41]. For the profiled analysis, the maximum likelihood tool (for example, template attacks) is the most efficient [38]. Thereby, all our following studies concern these attacks (CPA, LRA, and template attacks).

References

1. Spreitzer R, Moonsamy V, Korak T, Mangard S (2018) Systematic classification of side-channel attacks: a case study for mobile devices. IEEE Commun Surv Tutor 20(1):465–488
2. Kocher PC (1996) Timing attacks on implementations of Diffie-Hellman, RSA, DSS, and other systems. In: Koblitz N (ed) Advances in cryptology - CRYPTO'96, 16th annual international cryptology conference, Santa Barbara, California, USA, August 18–22, 1996, Proceedings. Lecture notes in computer science, vol 1109. Springer, pp 104–113
3. Kocher P, Jaffe J, Jun B (1999) Differential power analysis. In: Advances in cryptology - CRYPTO'99. Springer, pp 388–397
4. Messerges TS, Dabbish EA, Sloan RH (2002) Examining smart-card security under the threat of power analysis attacks. IEEE Trans Comput 51(5):541–552
5. Akkar M-L, Bevan R, Dischamp P, Moyart D (2000) Power analysis, what is now possible... In: ASIACRYPT. Lecture notes in computer science, vol 1976. Springer, pp 489–502
6. Le T-H, Clédière J, Canovas C, Robisson B, Servière C, Lacoume J-L (2006) A proposition for correlation power analysis enhancement. In: CHES, Yokohama, Japan. LNCS, vol 4249. Springer, pp 174–186
7. Pearson K (1900) On lines and planes of closest fit to points in space. Philos Mag 2:559–572
8. Coron J-S, Kocher PC, Naccache D (2000) Statistics and secret leakage. In: Financial cryptography, February 20–24 2000, Anguilla, British West Indies. Lecture notes in computer science, vol 1962. Springer, pp 157–173
9. Mayer-Sommer R (2001) Smartly analyzing the simplicity and the power of simple power analysis on smartcards. In: CHES, May 14–16 2001. LNCS, vol 1965. Springer, pp 78–92. http://citeseer.nj.nec.com/mayer-sommer01smartly.html
10. Brier É, Clavier C, Olivier F (2004) Correlation power analysis with a leakage model. In: Joye M, Quisquater J-J (eds) Cryptographic hardware and embedded systems - CHES 2004: 6th international workshop Cambridge, MA, USA, August 11–13, 2004. Proceedings. Lecture notes in computer science, vol 3156. Springer, pp 16–29

11. Bottinelli P, Bos JW (2017) Computational aspects of correlation power analysis. J Cryptogr Eng 7(3):167–181
12. Guillot P, Millérioux G, Dravie B, El Mrabet N (2017) Spectral approach for correlation power analysis. In: El Hajji S, Nitaj A, Souidi EM (eds) Codes, cryptology and information security - 2nd international conference, C2SI 2017, Rabat, Morocco, April 10–12, 2017, Proceedings - In: Honor of Claude Carlet. Lecture notes in computer science, vol 10194. Springer, pp 238–253
13. Hessc CA, Nortey F, Ofosu J (2018) Introduction to nonparametric statistic methods
14. Batina L, Gierlichs B, Lemke-Rust K (2008) Comparative evaluation of rank correlation based DPA on an AES prototype chip. In: ISC, September 15–18 2008, Taipei, Taiwan. Lecture notes in computer science, vol 5222. Springer, pp 341–354
15. Heuser A, Rioul O, Guilley S (2014) Good is not good enough - deriving optimal distinguishers from communication theory. In: Batina L, Robshaw M (eds) Cryptographic hardware and embedded systems - CHES 2014 - 16th international workshop, Busan, South Korea, September 23–26, 2014. Proceedings. Lecture notes in computer science, vol 8731. Springer, pp 55–74
16. Amir M, Oliver M, Thomas E (2010) Attack correlation-enhanced power analysis collision. In: CHES, August 17–20, 2010, Santa Barbara, CA, USA. Lecture notes in computer science, vol 6225. Springer, pp 125–139
17. Bogdanov A (2007) Improved side-channel collision attacks on AES. In: Adams CM, Miri A, Wiener MJ (eds) Selected areas in cryptography. Lecture notes in computer science, vol 4876. Springer, pp 84–95
18. Hanley N, Kim HS, Tunstall M (2015) Exploiting collisions in addition chain-based exponentiation algorithms using a single trace. In Nyberg K (ed) Topics in cryptology - CT-RSA 2015, the cryptographer's track at the RSA conference 2015, San Francisco, CA, USA, April 20–24, 2015. Proceedings. Lecture notes in computer science, vol 9048. Springer, pp 431–448
19. Bauer A, Jaulmes É, Prouff E, Reinhard J-R, Wild J (2015) Horizontal collision correlation attack on elliptic curves - extended version. Cryptogr Commun 7(1):91–119
20. Gierlichs B, Batina L, Tuyls P, Preneel B (2008) Mutual information analysis. In: CHES, 10th international workshop, August 10–13, 2008, Washington, D.C, USA. Lecture notes in computer science, vol 5154. Springer, pp 426–442
21. Batina L, Gierlichs B, Prouff E, Rivain M, Standaert F-X, Veyrat-Charvillon N (2011) Mutual information analysis: a comprehensive study. J Cryptol 24(2):269–291
22. Veyrat-Charvillon N, Standaert F-X (2009) Mutual information analysis: how, when and why? In: Clavier C, Gaj K (eds) Cryptographic hardware and embedded systems - CHES 2009, 11th international workshop, Lausanne, Switzerland, September 6–9, 2009, Proceedings. Lecture notes in computer science, vol 5747. Springer, pp 429–443
23. de Chérisey E, Guilley S, Rioul O, Piantanida P (2019) Best information is most successful mutual information and success rate in side-channel analysis. IACR Trans Cryptogr Hardw Embed Syst 2019(2):49–79
24. Standaert F-X, Malkin T, Yung M (2009) A unified framework for the analysis of side-channel key recovery attacks. In: EUROCRYPT, April 26–30 2009, Cologne, Germany. LNCS, vol 5479. Springer, pp 443–461
25. Benedikt G, Lejla B, Bart P, Ingrid V (2010) Revisiting higher-order DPA, attacks: multivariate mutual information analysis. In: CT-RSA, March 1–5 2010, San Francisco, CA, USA. LNCS, vol 5985. Springer, pp 221–234
26. Le T-H, Berthier M (2010) Mutual information analysis under the view of higher-order statistics. In: Echizen I, Kunihiro N, Sasaki R (eds) IWSEC. Lecture notes in computer science, vol 6434. Springer, pp 285–300
27. Moradi A, Mousavi N, Paar C, Salmasizadeh M (2009) A comparative study of mutual information analysis under a Gaussian assumption. In: WISA (information security applications, 10th international workshop), August 25–27 2009, Busan, Korea. Lecture Notes in Computer Science, vol 5932. Springer, pp 193–205

28. Prouff E, Rivain M (2010) Theoretical and practical aspects of mutual information-based side channel analysis. Int J Appl Cryptogr (IJACT) 2(2):121–138
29. Whitnall C, Oswald E (2011) A comprehensive evaluation of mutual information analysis using a fair evaluation framework. In: Rogaway P (ed) CRYPTO. Lecture notes in computer science, vol 6841. Springer, pp 316–334
30. de Chérisey E, Guilley S, Heuser A, Rioul O (2018) On the optimality and practicability of mutual information analysis in some scenarios. Cryptogr Commun 10(1):101–121
31. Whitnall C, Oswald E, Mather L (2011) An exploration of the Kolmogorov-Smirnov test as a competitor to mutual information analysis. In Prouff E (ed) CARDIS. Lecture notes in computer science, vol 7079. Springer, pp 234–251
32. Richter B, Knichel K, Moradi A (2019) A comparison of χ^2-test and mutual information as distinguisher for side-channel analysis. In: Belaïd S, Güneysu T (eds) Smart card research and advanced applications - 18th international conference, CARDIS 2019, Prague, Czech Republic, November 11–13, 2019, Revised selected papers. Lecture notes in computer science, vol 11833. Springer, pp 237–251
33. Toth R, Faigl Z, Szalay M, Imre S (2008) An advanced timing attack scheme on RSA. In: Telecommunications network strategy and planning symposium, 2008. Networks 2008. The 13th international, pp 1–24
34. Anderson TW (1962) On the distribution of the two-sample Cramer-von Mises criterion. Ann Math Stat 33(3):1148–1159
35. Goodwill G, Jun B, Jaffe J, Rohatgi P (2011) A testing methodology for side-channel resistance validation, September 2011. NIST non-invasive attack testing workshop, http://csrc.nist.gov/news_events/non-invasive-attack-testing-workshop/papers/08_Goodwill.pdf
36. Tobias S, Amir M, Tim G (2015) Arithmetic addition over Boolean masking - towards first- and second-order resistance in hardware. IACR Cryptol ePrint Arch 2015:66
37. Moradi A, Richter B, Schneider T, Standaert F-X (2018) Leakage detection with the x2-test. IACR Trans Cryptogr Hardw Embed Syst 2018(1):209–237 Feb
38. Chari S, Rao JR, Rohatgi P (2002) Template attacks. In: Kaliski BS, Jr, Koç ÇK, Paar C (eds) Cryptographic hardware and embedded systems - CHES 2002, 4th international workshop, Redwood Shores, CA, USA, August 13–15, 2002, Revised papers. Lecture notes in computer science, vol 2523. Springer, pp 13–28
39. Stjepan P, Annelie H, Sylvain G (2017) Template attack versus Bayes classifier. J Cryptogr Eng 7:09
40. Werner S (2008) Advanced stochastic methods in side channel analysis on block ciphers in the presence of masking. J Math Cryptol 2(3), 291–310, ISSN (Online) 1862–2984. ISSN (Print) 1862–2976. https://doi.org/10.1515/JMC.2008.013
41. Schindler W (2005) On the optimization of side-channel attacks by advanced stochastic methods. In: Vaudenay S (ed) Public key cryptography - PKC 2005, 8th international workshop on theory and practice in public key cryptography, Les Diablerets, Switzerland, January 23–26, 2005, Proceedings. Lecture notes in computer science, vol 3386. Springer, pp 85–103
42. Doget J, Prouff E, Rivain M, Standaert F-X (2011) Univariate side channel attacks and leakage modeling. J Cryptogr Eng 1(2):123–144
43. Heuser A, Zohner M (2012) Intelligent machine homicide - breaking cryptographic devices using support vector machines. In: Schindler W, Huss SA (eds) COSADE. LNCS, vol 7275. Springer, pp 249–264
44. Markowitch O, Lerman L, Bontempi G (2011) Side channel attack: an approach based on machine learning
45. Hettwer B, Gehrer S, Güneysu T (2019) Applications of machine learning techniques in side-channel attacks: a survey. J Cryptogr Eng
46. Maghrebi H, Portigliatti T, Prouff E (2016) Breaking cryptographic implementations using deep learning techniques. In Carlet C, Hasan MA, Saraswat V (eds) Security, privacy, and applied cryptography engineering - 6th international conference, SPACE 2016, Hyderabad,

India, December 14–18, 2016, Proceedings. Lecture notes in computer science, vol 10076. Springer, pp 3–26

47. Cagli E, Dumas C, Prouff E (2017) Convolutional neural networks with data augmentation against jitter-based countermeasures - profiling attacks without pre-processing. In: Fischer W, Homma N (eds) Cryptographic hardware and embedded systems - CHES 2017 - 19th international conference, Taipei, Taiwan, September 25–28, 2017, Proceedings. Lecture notes in computer science, vol 10529. Springer, pp 45–68

48. Timon B (2019) Non-profiled deep learning-based side-channel attacks with sensitivity analysis. IACR Trans Cryptogr Hardw Embed Syst 2019(2):107–131 Feb

49. Standaert FX, Peeters E, Quisquater J-J (2005) On the masking countermeasure and higher-order power analysis attacks. In: International conference on information technology: coding and computing, 2005. ITCC 2005, vol 1, pp 562–567

50. Schramm K, Paar C (2006) Higher order masking of the AES. In: Pointcheval D (ed) CT-RSA. LNCS, vol 3860. Springer, pp 208–225

51. Joye M, Paillier P, Schoenmakers B (2005) On second-order differential power analysis. In: CHES, August 29 – September 1st 2005, Edinburgh, UK. LNCS, vol 3659. Springer, pp 293–308

52. Oswald E, Mangard S (2007) Template attacks on masking — resistance is futile. In: Abe M (ed) CT-RSA. Lecture notes in computer science, vol 4377. Springer, pp 243–256

53. Prouff E, Rivain M, Bevan R (2009) Statistical analysis of second order differential power analysis. IEEE Trans Comput 58(6):799–811

54. Standaert F-X, Veyrat-Charvillon N, Oswald E, Gierlichs B, Medwed M, Kasper M, Mangard S (2010) The world is not enough: another look on second-order DPA. In: ASIACRYPT, December 5–9 2010, Singapore. LNCS, vol 6477. Springer, pp 112–129. http://www.dice.ucl.ac.be/~fstandae/PUBLIS/88.pdf

55. Bruneau N, Guilley S, Heuser A, Rioul O (2014) Masks will fall off – higher-order optimal distinguishers. In: Sarkar P, Iwata T (eds) Advances in cryptology – ASIACRYPT 2014 - 20th international conference on the theory and application of cryptology and information security, Kaoshiung, Taiwan, R.O.C., December 7–11, 2014, Proceedings, Part II. Lecture notes in computer science, vol 8874. Springer, pp 344–365

56. Bruneau N, Guilley S, Heuser A, Rioul O (2014) Masks will fall off: higher-order optimal distinguishers. In: ASIACRYPT, December 2014. LNCS, vol 8874. Springer, pp 344–365. Sarkar P, Iwata T (eds) ASIACRYPT 2014, PART II

57. Lomné V, Prouff E, Rivain M, Roche T, Thillard A (2014) How to estimate the success rate of higher-order side-channel attacks. In: Batina L, Robshaw M (eds) Cryptographic hardware and embedded systems - CHES 2014 - 16th international workshop, Busan, South Korea, September 23–26, 2014. Proceedings. Lecture notes in computer science, vol 8731. Springer, pp 35–54

58. Mangard S, Oswald E, Standaert F-X (2011) One for all - all for one: unifying standard differential power analysis attacks. IET Inf Secur 5(2):100–110

59. Whitnall C, Oswald E (2011) A fair evaluation framework for comparing side-channel distinguishers. J Cryptogr Eng 1(2):145–160

60. de Chérisey É, Guilley S, Heuser A, Rioul O (2017) On the optimality and practicability of mutual information analysis in some scenarios. Cryptogr Commun

61. Dabosville G, Doget J, Prouff E (2013) A new second-order side channel attack based on linear regression. IEEE Trans Comput 62(8):1629–1640

SCA Countermeasures

4

SCAs succeed because the leakage of cryptographic devices depends on the sensitive variables. Consequently, all the countermeasures aim at breaking this dependency, or at least to reduce it. Essentially they boil down to randomizing the leakage, or to making it equal in each device clock cycle. Subsequently, there are two main countermeasure classes, namely, *hiding* and *masking* (see [1]). Those countermeasures are summarized in this chapter.

In the rest of the book, hiding countermeasures can be assimilated to the study of SCA when the noise signal-to-noise ratio (SNR) is small, whereas masking requires specific processing (addressed in Sects. 6.4, 8.4, and 10).

4.1 Hiding

Hiding aims at breaking the link between the leakage of the cryptographic devices and the processed data values. In practice, the hiding countermeasures maintain unchanged the way the algorithms execute (compared to unprotected devices), by computing on the same sensitive values. But hiding makes it difficult to find exploitable information in the leakage. Concretely, the ideal hiding cannot be reached. But many pragmatic methods are suggested to reduce the link between the leakage and the sensitive variables. These methods are divided into two categories. In the first one, the leakage is randomized by executing the sensitive operations at different instants from an execution to another. In the second one, the methods do not affect the time dimension of the leakage, but they affect its amplitude dimension [2, Chaps. 7, 8].

© The Author(s), under exclusive license to Springer Nature Switzerland AG 2021
M. Ouladj and S. Guilley, *Side-Channel Analysis of Embedded Systems*,
https://doi.org/10.1007/978-3-030-77222-2_4

4.1.1 Hiding on the Time Dimension

During the attack, the traces should be aligned first. In fact, if this condition does not hold, then the SCA is significantly affected (it needs more traces). Consequentially, cryptographic designers randomize the execution of the cryptographic algorithm control flow without changing the algorithm itself. In practice, the random insertion of dummy operations, jittering, and the shuffling of operations is the most used technique to randomize the cryptographic algorithm execution in the time dimension.

4.1.1.1 Random Insertion of Dummy Operations

The added dummy operations should not be distinguished for real ones. In practice, adding simple *nop* instructions could be enough. One can also randomly repeat the genuine operations but with dummy values [3].

4.1.1.2 Jittering Countermeasure

It is a countermeasure that consists in de-synchronizing the traces by adding a random waiting time between the execution of the instructions. A disadvantage of this countermeasure is the required large number of random numbers (the delays) to be generated. Furthermore, a Convolutional Neural Network (CNN)-based SCA can easily overcome this countermeasure [4].

4.1.1.3 Shuffling of Operations

Shuffling of operations consists in executing the independent sensitive operations in a different order for each run of the cryptographic algorithm [3]. Shuffling reduces the height of the correlation peaks. For example, if a shuffling scheme randomizes uniformly the advent of two sensitive variables, then the amplitude of the correlation peak is reduced by a factor 2, when one attacks both of the sensitive variables [2, Chap. 10]. This effect holds for higher order DPA attacks, Sect. 3.3.

For more degrees of freedom in the randomization, the insertion of dummy operations can be used alongside with the jittering and the shuffling techniques. In such scheme shuffling spreads the leakage containing the information about a sensitive variable into several signals [5]. Only one of them comes from the genuine operation.

4.1.2 Hiding on the Amplitude Dimension

Hiding the secret can also be achieved on the amplitude dimension of the leakage. The countermeasure, therefore, affects the leakage characteristic instead of its appearance in time. It can be implemented either by randomizing the amplitude or by making it constant. In fact, the success rate of SCAs depends on the SNR of the leakage [6]. Since the higher the SNR the higher the SCAs success rate, designers aim to reduce the SNR either by increasing the noise or by reducing the signal.

4.1.2.1 Increasing the Noise

It can be achieved by (1) performing several independent operations in parallel with the sensitive one or (2) by using dedicated noise engines that perform random switching in parallel [2, Chap. 7].

4.1.2.2 Reducing the Signal

To be accurate, reducing the signal amounts to reducing its variance. It is theoretically achieved by making it constant. In practice, there exist two techniques to make it quasi-constant. The first one consists in filtering the power consumption by removing all its data-dependent and operation-dependent components. The second technique consists in employing dedicated logic styles for the cells of cryptographic circuits. It is typically implemented as dual-rail precharge (DRP) logic styles [7–9]. Recently, authors of [10] show that it holds only for power analysis. In contrast, using high-resolution electromagnetic analysis, they prove that it cannot hide from localized EM.

4.2 Masking Countermeasure

Masking a countermeasure aims at breaking the dependency between all sensitive variable Z and the corresponded leakage, or at least to reduce it, by randomizing the sensitive variables during its processing. An advantage of this approach is its ability to be implemented at the algorithm level without changing the leaking characteristics of the cryptographic device [2, Chap. 9]. Furthermore, it is a *provably secure* countermeasure.

To define the masking formally, let d be a strictly positive integer. Masking consists in splitting randomly each sensitive value $z = F(x, k^*)$ to $d + 1$ values (called shares) denoted by $F_R^{(w)}(x, k^*) = f^{(w)}(x, k^*, r^{(w)})$ such that $w \in \{0, \dots, d\}$ and $r^{(w)}$ denote random values generated internally afresh for each execution. For the sake of clarity, perfect higher order masking shall be formally defined as follows.

Definition 1 (*Perfect dth-order masking* [11]) Let R be a random variable. Let us denote the random variables $F_R^{(w)}(x, k) = f^{(w)}(x, k, R^{(w)})$ for $w \in \{0, \dots, d\}$ and for all possible pair (x, k). A masking scheme is perfect at dth-order if the joint distribution of maximum d shares (from $\{F_R^{(0)}(x, k), \dots, F_R^{(d)}(x, k)\}$) is independent from the pair (x, k).

It is notable that dth-order security implies 1st, 2nd, $\dots$, $(d-1)$th-order security.

The advantage of the (perfect) masking is that the required number of messages required to succeed attacks increases exponentially according to the number of shares d [12]. In contrast, we will show in the sequel that the complexity of the masked implementation often increases linearly according to d.

In practice, for each sensitive variable Z the device generates d random variables $R^{(1)} \cdots R^{(d)}$ internally and instead of computing Z it computes $Z * R^{(1)} * \cdots * R^{(d)}$ (where $*$ denotes any operation chosen by the designers). It is noteworthy that the masks are freshly generated and (re)used as little as possible. Of course, designers must retrieve the correct ciphertext (or plaintext) at the end of the algorithm, using $Z * R^{(1)} * \cdots * R^{(d)}$ (called masked value in the sequel) and the random variables $R^{(1)} \cdots R^{(d)}$ (called masks).

To retrieve the correct ciphertext (resp. plaintext) at the end of ciphering (resp. deciphering), the designers must follow the masks' effect for each internal transformation of the current sensitive variable. Formally, for each internal transformation $f : E \longrightarrow E$, the designers can (for example) define a masking transformation $mask : E \longrightarrow F$ and a modified transformation $f' : F \longrightarrow F$, such that

$$f'(mask(z)) = mask(f(z)). \tag{4.1}$$

Then, during the implementation, designers replace the composite transformation "$mask \circ f$" by "$f' \circ mask$". Of course, the transformation f' must not reveal any information about z.

Since the discovery of SCAs, several masking schemes have been introduced in the literature.

4.2.1 Boolean Masking (BM)

It is the first masking scheme to defeat the (first-order) SCA. It consists in splitting each sensitive variable Z to the following $d+1$ variables: $Z^{(0)} = Z \oplus R^{(1)} \oplus \cdots \oplus R^{(d)}$ (the masked variable) and the d random variables $Z^{(1)} = R^{(1)}, ..., Z^{(d)} = R^{(d)}$ (the masks) [13,14]. This means that $Z = \oplus_{i=0}^{d} Z^{(i)}$, and any strict subset of the shares' set $\{Z^{(1)} \cdots Z^{(d)}\}$ do not reveal any information about Z.

This masking scheme is $\mathbb{F}_2$-linear. This means for any z, z' in E, $mask(z \oplus z') = mask(z) \oplus mask(z')$. Furthermore, if the internal transformation f is linear, then $mask(f(z \oplus z')) = mask(f(z)) \oplus mask(f(z'))$. So, to efficiently mask $f(z \oplus z')$, it is sufficient to mask separately $f(z)$ and $f(z')$, and then compute the exclusive-or $mask(f(z)) \oplus mask(f(z'))$.

If f is non-linear (as an S-box of AES or a multiplication) the implementation of masking is not trivial. Thereby, several other schemes have been introduced in the literature.

4.2.1.1 Ishai-Sahai-Wagner's Scheme (ISW)

Ishai et al. introduce the first masking scheme to protect non-linear transformation for the Boolean masking [15]. Furthermore, they introduce the *probing model/security* concept, which has kicked off the work on proven secure masking schemes. The ISW scheme consists in transforming each operation of the cryptographic device into a combination of several "NOT" and "AND" Boolean operations between binary variables.

Similar to above, the NOT operation is linear (NOT(z) = $z \oplus 1$, for z in $\mathbb{F}_2$), so its Boolean masking is trivial. In contrast, to mask an $\mathbb{F}_2$-product AND(z, z') = zz' at the first order, Ishai et al. provide an algorithm that requires $(d + 1)^2$ multiplications (AND), $2d(d + 1)$ additions (XOR), and $d(d + 1)/2$ generated random values.

4.2.1.2   Rivain and Prouff's Scheme (RP)

In order to protect the AES implementation, using Boolean masking, Rivain and Prouff extend the ISW scheme from $\mathbb{F}_2$ to $\mathbb{F}_2^n$ (where n is typically equal to 8) [16]. Recall that $\mathbb{F}_2^n$ and the AES polynomials' field $\mathbb{F}_2[\alpha]/\langle P(\alpha) \rangle$ are isomorphic. Even if their multiplication algorithm has also a quadratic complexity in d, they provide efficient algorithms for exponentiation and inversion computations. They benefit from the linear complexity of the exponentiation by any power of 2. Subsequently, several improvements and generalizations of this scheme have been suggested, as in [17].

4.2.2   Multiplicative Masking (MM)

Multiplicative masking consists in splitting all the sensitive variables in a multiplicative way instead of the additive ($\oplus$) operation. It is more adapted for the computation of some cryptographic function (such as the AES S-box [18]). But it has to be implemented carefully in order to prevent the first-order DPA with a zero-value power model [19].

4.2.3   Affine Masking (AfM)

Affine masking is firstly introduced by von Willich in 2001 [20] as an alternative to the Boolean masking. It consists in masking each sensitive value $z \in \mathbb{F}_2^n$ into $mask(z) = r^{(0)} \cdot z \oplus r^{(1)}$, where $r^{(0)}$ and $r^{(1)}$ are random masks. In fact, it leads to a significant gain in terms of security compared to the multiplicative and the Boolean masking [21], but it requires the generation of a larger number of random numbers.

4.2.4   Arithmetic Masking (ArM)

If the cipher algorithm involves an arithmetic operation such as a modular addition and/or a modular multiplication (as in IDEA, RC6, and the HMAC Construction), then it is useful to use the corresponded arithmetic operation as a masking scheme instead of the Boolean operations [22,23]. In fact, if the cipher algorithm involves a mix of Boolean and arithmetic operations, then a secure switching between the Boolean and the arithmetic masking must be managed [24–29].

4.2.5 Polynomials-Based Masking (PM)

The polynomials-based masking was independently introduced by Goubin and Martinelli in [30] on the one hand, and Prouff and Roche in [31] on the other hand. Both of them use Shamir's Secret Sharing (SSS) [32] for sharing the sensitive data. In fact, SSS is very closely related to the Maximum Distance Separable (MDS) codes like Reed-Solomon codes [33]. This scheme is a generalization of the Boolean masking and also adapted to the linear transformations, but not for the non-linear ones. Thereby, for the multiplication, Goubin and Martinelli suggest the use of the so-called BGW secure scheme of multiparty computation [34]. In contrast, Prouff and Roche provide their own multiplication algorithm [31].

4.2.6 Leakage Squeezing Masking (LSM)

In the LSM scheme, the shares are formed as for perfect masking (Boolean masking with all possible values for the masks) except that some bijective functions $(F_w)_{w=1,\ldots,d}$ are applied to the masks [35,36], such that the masked sensitive value z_0 is

$$z_0 = z \oplus F_1(m^{(1)}) \oplus \cdots \oplus F_d(m^{(d)}).$$

That leads to a better mixing of bits, as discussed and demonstrated in [37–39].

4.2.7 Rotating S-Boxes Masking (RSM)

For a security/complexity compromise in the masking of the AES S-boxes, authors of [40] introduce the Rotating S-boxes Masking (RSM) [41,42] countermeasure. This scheme is the Boolean masking, where the masks are chosen uniformly from a code [38] of length $n = 8$ in $\mathbb{F}_2$, either:

- $\mathbb{C}_0 = \{0\text{x}00\}$ (no masking);
- $\mathbb{C}_1 = \{0\text{x}00, 0\text{xff}\}$;
- $\mathbb{C}_2$, a non-linear code of length 8, size 12, and its generating matrix denoted G; or
- $\mathbb{C}_3$ is a linear code of length 8, and dimension 4 (see paper [43] for more details).

The last case is interesting since there are 16 masks (i.e., 16 masked S-boxes' tables) [38,44,45], which matches the number of bytes in the datapath of AES.

4.2.8 Inner Product Masking (IPM)

The IPM countermeasure is introduced in [46–48] to generalize the Boolean masking. It consists in splitting each sensitive value z into $d + 1$ random values $z^{(0)} \cdots z^{(d)}$, not such as $z = \oplus_{i=0}^{d} z^{(i)}$ as in Boolean masking but such as $z = \oplus_{i=0}^{d} u^{(i)} z^{(i)}$, where $u^{(i)}$ are public values in $\mathbb{F}_{2^n} \backslash \{0\}$. It is called inner product masking, because it can be seen as $z = \langle (u^{(0)} \cdots u^{(d)}); (z^{(0)} \cdots z^{(d)}) \rangle$. Thanks to the bi-linearity of the inner product, the complexity of IPM for any linear transformation is linear on d. For the multiplication masking, the interested reader should refer to [49].

It is shown in [46] that IPM is a generalization of both Boolean Sect. 4.2.1, multiplicative Sect. 4.2.2, affine Sect. 4.2.3, and polynomial masking Sect. 4.2.5. Subsequently, several studies of this scheme have been leading as in [49,50].

More recently, an IPM with Fault Detection scheme (IPM-FD) has been introduced to prevent both SCA and fault injection attacks [51]. It is notable that there is an insistent recommendation for dual side-channel (SCA) and fault injection (FIA) protections. Thereby, the so-called Direct Sum Masking (ODSM) is introduced in the sequel.

4.2.9 Direct Sum Masking (DSM)

The DSM countermeasure is an error correction code-based masking scheme introduced in [52,53] that provides a dual protection (against SCA and FIA). Unlike IPM (which is defined over $\mathbb{F}_{2^n}$), DSM is defined at bit level (i.e., in $\mathbb{F}_2$). To describe the DSM scheme, let us see the sensitive value z (resp. the mask r) as a bitvector $z = (z_1, z_2, \ldots z_n)$ (resp. $r = (r_1, r_2, \ldots r_{n'})$). The DSM scheme consists in encoding the sensitive value in an n-dimensional code $\mathscr{C}$ and encoding the mask in an n'-dimensional code $\mathscr{D}$, such that $\mathscr{C}$ and $\mathscr{D}$ are complementary (i.e., $\mathbb{F}_2^{n+n'} = \mathscr{C} \oplus \mathscr{D}$). So, masking z by r means adding (XOR operation) the encoded word of r to that of z. Owing to this useful structure, the designers can easily extract the sensitive value from the masked one (by projection). The discussion from [52] requires the code $\mathscr{C}$ and $\mathscr{D}$ to be dual. This scheme is consequently called Orthogonal Direct Sum Masking (ODSM).

IPM is a specific case of DSM. Namely, IPM can be written as a DSM by adapting the DSM scheme base field from $\mathbb{F}_2^{n+n'}$ to $\mathbb{F}_{2^{n+n'}}$ [54]. A DSM scheme written over the space $\mathbb{F}_2[\alpha]/P(\alpha)$, such that the degree of P equals n (instead of its isomorphic space $\mathbb{F}_{2^n}$), is called Polynomial Direct Sum Masking (PDSM) [55]. Advantageously, the PDSM countermeasure is adapted for the AES masking when the polynomial for the countermeasure is the same as the AES defining polynomial $\mathbb{F}_2[\alpha]/P(\alpha)$.

It is shown in the literature that the DSM protects the implementation against any $\{d_\mathscr{C} - 1\}$-order SCA, where $d_\mathscr{C}$ is the minimum distance of the code $\mathscr{C}$ [55].

4.2.9.1 Multi-shares Direct Sum Masking (DSM)

The extension of DSM to a Multi-Shares version (MS-DSM) will be studied in Appendix 10.7. In this appendix, we shall explain how to extend DSM (introduced in Sect. 4.2.9) to a multi-share case. This appendix is divided into two subsections: how not to extend to an incorrect multi-share DSM (Sect. 10.7.1) and how it should be correctly extended (Sect. 10.7.2). Both methods are supported by arguments based on coding theory.

4.2.10   Comparison Between Masking Schemes

The main advantage of the simpler masking schemes, such as additive (with respect to XOR) and the multiplicative ones, is its efficient possible implementations [56]. In contrast, the more recent schemes (polynomial masking, IPM, and DMS) are more complex and subsequently their implementations are more expensive [57]. But, this complexity leads to improve the security of the masking [30,49,52]. Finally, let us mention that it is still an open problem to find the best trade-off between the implementation complexity and the masking security [54].

Random Number Generator Problem in masking: To efficiently generate the masks, authors of [58] introduce the notion of robust Pseudo-random Number Generator (PRNG) instead of the True one (TRNG). Against a t-probes adversary, PRNG requires $\tilde{O}(t^4)$ random bits. Using multiple independent PRGs instead of a single one, Coron et al. [59] suggest a practical construction of a robust PRNG, requiring only $\tilde{O}(t^2)$ random bits. Furthermore, the running time goes down from $\tilde{O}(t^4)$ to $\tilde{O}(t)$.

4.3   Combination of Countermeasures

For a good protection of the cryptographic implementations, one should combine different countermeasures at once. For example, this can be achieved using both masking, shuffling, and dummy operations as in [60] and [61], or combining additive and multiplicative secret sharing into an affine masking scheme that is additionally mixed with a shuffled execution as in [62].

In the rest of the book, we will study several new improvements of the SCAs implementation, resulting from a recent spectral approach of computations (Part II) combined with a new SCA's principle called Coalescence (Part III).

References

1. Guilley S, Meynard O, Nassar M, Duc G, Hoogvorst P, Maghrebi H, Elaabid A, Bhasin S, Souissi Y, Debande N, Sauvage L, Danger J-L (2011) Vade mecum on side-channels attacks and countermeasures for the designer and the evaluator. In: DTIS (design & technologies of integrated systems), IEEE, March 6–8 2011, Athens, Greece. IEEE. https://doi.org/10.1109/DTIS.2011.5941419; Online version: http://hal.archives-ouvertes.fr/hal-00579020/en/
2. Mangard S, Oswald E, Popp T (2006) Power analysis attacks: revealing the secrets of smart cards. Springer. http://www.springer.com/. ISBN 0-387-30857-1, http://www.dpabook.org/
3. Tillich S, Herbst C, Mangard S (2007) Protecting AES software implementations on 32-bit processors against power analysis. In: ACNS

4. Cagli E, Dumas C, Prouff E (2017) Convolutional neural networks with data augmentation against jitter-based countermeasures - profiling attacks without pre-processing. In: Fischer W, Homma N (eds) Cryptographic hardware and embedded systems - CHES 2017 - 19th international conference, Taipei, Taiwan, September 25–28, 2017, Proceedings. Lecture notes in computer science, vol 10529. Springer, pp 45–68

5. Rivain M, Prouff E, Doget J (2009) Higher-order masking and shuffling for software implementations of block ciphers. In: CHES, September 6–9 2009, Lausanne, Switzerland. Lecture notes in computer science, vol 5747. Springer, pp 171–188

6. Standaert F-X, Malkin T, Yung M (2009) A unified framework for the analysis of side-channel key recovery attacks. In: EUROCRYPT. LNCS, April 26–30 2009, Cologne, Germany, vol 5479. Springer, pp 443–461

7. Sokolov D, Murphy J, Bystrov AV, Yakovlev A (2004) Improving the security of dual-rail circuits. In CHES. LNCS, August 11–13 2004, Cambridge, MA, USA, vol 3156. Springer, pp. 282–297

8. Tiri K, Verbauwhede I (2004) A logic level design methodology for a secure DPA resistant ASIC or FPGA implementation. In: DATE'04, pp. 246–251. IEEE computer society (February 2004) Paris. France. https://doi.org/10.1109/DATE.2004.1268856

9. Tiri K, Akmal M, Verbauwhede I (2002) A dynamic and differential CMOS logic with signal independent power consumption to withstand differential power analysis on smart cards. In: European solid-state circuits conference (ESSCIRC), September 2002, Florence, Italy, pp 403–406. http://citeseer.ist.psu.edu/tiri02dynamic.html

10. Immler V, Specht R, Unterstein F (2018) Your rails cannot hide from localized EM: how dual-rail logic fails on FPGAS - extended version. J Cryptogr Eng 8(2):125–139

11. Blömer J, Guajardo J, Krummel V (2004) Provably secure masking of AES. In: LNCS, Proceedings of SAC'04, vol 3357, August 2004, Waterloo, Canada. Springer, pp 69–83

12. Prouff E, Rivain M (2013) Masking against side-channel attacks: a formal security proof. In: Johansson T, Nguyen PQ (eds) Advances in cryptology - EUROCRYPT 2013, 32nd annual international conference on the theory and applications of cryptographic techniques, Athens, Greece, May 26–30, 2013. Proceedings. Lecture notes in computer science, vol 7881. Springer, pp 142–159

13. Chari S, Jutla CS, Rao JR, Rohatgi P (1999) Towards sound approaches to counteract power-analysis attacks. In: Wiener MJ (ed) CRYPTO. Lecture notes in computer science, vol 1666. Springer, pp 398–412

14. Goubin L, Patarin J (1999) DES and differential power analysis. In: Koç ÇK, Paar C (eds) Cryptographic hardware and embedded systems. LNCS, vol 1717. Springer, pp 158–172

15. Yuval I, Amit S, David W (2003) Private circuits: securing hardware against probing attacks. In: CRYPTO. Lecture notes in computer science, vol 2729, August 17–21, 2003, Santa Barbara, California, USA. Springer, pp 463–481

16. Rivain M, Prouff E (2010) Provably secure higher-order masking of AES. In: Mangard S, Standaert F-X (eds) CHES. LNCS, vol 6225. Springer, pp 413–427

17. Goudarzi D, Rivain M, Vergnaud D, Vivek S (2017) Generalized polynomial decomposition for s-boxes with application to side-channel countermeasures. In: Fischer W, Homma N (eds) Cryptographic hardware and embedded systems - CHES 2017 - 19th international conference, Taipei, Taiwan, September 25–28, 2017, Proceedings. Lecture notes in computer science, vol 10529. Springer, pp 154–171

18. NIST/ITL/CSD. Advanced encryption standard (AES). FIPS PUB 197, Nov 2001. http://nvlpubs.nist.gov/nistpubs/FIPS/NIST.FIPS.197.pdf (also ISO/IEC 18033-3:2010)

19. De Meyer L, Reparaz O, Bilgin B (2018) Multiplicative masking for AES in hardware. IACR Trans Cryptogr Hardw Embed Syst 2018(3):431–468

20. Willich M (2001) A technique with an information-theoretic basis for protecting secret data from differential power attacks, pp 44–62

21. Fumaroli G, Martinelli A, Prouff E, Rivain M (2010) Affine masking against higher-order side channel analysis. In: Biryukov A, Gong G, Stinson DR (eds) Selected areas in cryptography. Lecture notes in computer science, vol 6544. Springer, pp 262–280
22. Jean-Sébastien C, Louis G (2000) Boolean on, masking arithmetic, against differential power analysis. In: CHES, August 17–18, 2000, Worcester, MA, USA. Lecture notes in computer science, vol 1965. Springer, pp 231–237
23. Kerstin L, Kai S, Paar C (2004) DPA, on n-bit sized boolean and arithmetic operations and its application to IDEA, RC6, and the HMAC-construction. In: CHES. Lecture notes in computer science, vol 3156, August 11–13, 2004, Cambridge, MA, USA. Springer, pp 205–219
24. Bettale L, Coron J-S, Zeitoun R (2018) Improved high-order conversion from Boolean to arithmetic masking. IACR Trans Cryptogr Hardw Embed Syst 2018(2):22–45 May
25. Hutter M, Tunstall M (2016) Constant-time higher-order Boolean-to-arithmetic masking. J Cryptogr Eng 9:173–184
26. Coron J-S, Großschädl J, Vadnala PK (2014) Secure conversion between Boolean and arithmetic masking of any order. In: Batina L, Robshaw M (eds) Cryptographic hardware and embedded systems - CHES 2014 - 16th international workshop, Busan, South Korea, September 23–26, 2014. Proceedings. Lecture notes in computer science, vol 8731. Springer, pp 188–205
27. Vadnala PK, Großschädl J (2013) Algorithms for switching between Boolean and arithmetic masking of second order. In: Gierlichs B, Guilley S, Mukhopadhyay D (eds) Security, privacy, and applied cryptography engineering - 3rd international conference, SPACE 2013, Kharagpur, India, October 19–23, 2013. Proceedings. Lecture notes in computer science, vol 8204. Springer, pp 95–110
28. Debraize B (2012) Efficient and provably secure methods for switching from arithmetic to Boolean masking. In: Prouff E, Schaumont P (eds) CHES. LNCS, vol 7428. Springer, pp 107–121
29. Goubin L (2001) A sound method for switching between Boolean and arithmetic masking. In: Koç ÇK, Naccache D, Paar C (eds) CHES. Lecture notes in computer science, vol 2162. Springer, pp 3–15
30. Goubin L, Martinelli A (2011) Protecting AES with Shamir's secret sharing scheme. In: Preneel B, Takagi T (eds) Cryptographic hardware and embedded systems - CHES 2011 - 13th international workshop, Nara, Japan, September 28 - October 1, 2011. Proceedings. LNCS, vol 6917. Springer, pp 79–94
31. Prouff E, Roche T (2011) Higher-Order Glitches Free Implementation of the AES Using Secure Multi-party Computation Protocols. In: Preneel B, Takagi T (eds) Cryptographic hardware and embedded systems - CHES 2011 - 13th international workshop, Nara, Japan, September 28 - October 1, 2011. Proceedings. LNCS, vol 6917. Springer, pp 63–78
32. Shamir A (1979) How to share a secret. Commun ACM 22(11):612–613 November
33. McEliece RJ, Sarwate DV (1981) On sharing secrets and Reed-Solomon codes. Commun ACM 24(9):583–584 September
34. Ben-Or M, Goldwasser S, Wigderson A (1988) Completeness theorems for non-cryptographic fault-tolerant distributed computation (extended abstract), pp 1–10
35. Carlet C, Danger J-L, Guilley S, Maghrebi H (2012) Leakage squeezing of order two. In: Galbraith SD, Nandi M (eds) Progress in cryptology - INDOCRYPT 2012, 13th international conference on cryptology in India, Kolkata, India, December 9–12, 2012. Proceedings. Lecture notes in computer science, vol 7668. Springer, pp 120–139
36. Karmakar S, Chowdhury DR (2013) Leakage squeezing using cellular automata. In: Kari J, Kutrib M, Malcher A (eds) Automata. Lecture Notes in Computer Science, vol 8155. Springer, pp 98–109
37. Carlet C, Danger J-L, Guilley S, Maghrebi H (2014) Leakage squeezing: optimal implementation and security evaluation. J Math Cryptol 8(3):249–295
38. Guilley S, Heuser A, Rioul O (2017) Codes for side-channel attacks and protections. In: El Hajji S, Nitaj A, Souidi EM (eds) Codes, cryptology and information security - 2nd international

conference, C2SI 2017, Rabat, Morocco, April 10–12, 2017, Proceedings - in honor of Claude Carlet. Lecture notes in computer science, vol 10194. Springer, pp 35–55

39. Danger J-L, Guilley S (2009) Protection des modules de cryptographie contre les attaques en observation d'ordre élevé sur les implémentations à base de masquage, 20 Janvier 2009. Brevet Français FR09/50341, assigné à l'Institut TELECOM

40. Nassar M, Souissi Y, Guilley S, Danger J-L (2012) RSM: a small and fast countermeasure for AES, secure against 1st and 2nd-order zero-offset SCAs. In: Rosenstiel W, Thiele L (eds) 2012 design, automation & test in Europe conference & exhibition, DATE 2012, Dresden, Germany, March 12–16, 2012. IEEE, pp 1173–1178

41. Kutzner S, Poschmann A (2014) On the security of RSM - presenting 5 first- and second-order attacks. In: Prouff E (ed) constructive side-channel analysis and secure design - 5th international workshop, COSADE 2014, Paris, France, April 13–15, 2014. Revised selected papers. Lecture notes in computer science, vol 8622. Springer, pp 299–312

42. Yamashita N, Minematsu K, Okamura T, Tsunoo Y (2014) A smaller and faster variant of RSM. In: DATE. IEEE, pp 1–6

43. Carlet C, Guilley S (2013) Side-channel indistinguishability. In: HASP, New York, NY, USA, June 23–24 2013. ACM, pp 9:1–9:8

44. Carlet C (2013) Correlation-immune Boolean functions for leakage squeezing and rotating S-box masking against side channel attacks. In: Gierlichs B, Guilley S, Mukhopadhyay D (eds) Security, privacy, and applied cryptography engineering - 3rd international conference, SPACE 2013, Kharagpur, India, October 19–23, 2013. Proceedings. Lecture notes in computer science, vol 8204. Springer, pp 70–74

45. DeTrano A, Karimi N, Karri R, Guo X, Carlet C, Guilley S (2015) Exploiting small leakages in masks to turn a second-order attack into a first-order attack and improved rotating substitution box masking with linear code cosets. Sci World J 10. https://doi.org/10.1155/2015/743618.

46. Balasch J, Faust S, Gierlichs B, Verbauwhede I (2012) Theory and practice of a leakage resilient masking scheme. In: Wang X, Sako K (eds) Advances in cryptology - ASIACRYPT 2012 - 18th international conference on the theory and application of cryptology and information security, Beijing, China, December 2–6, 2012. Proceedings. Lecture notes in computer science, vol 7658. Springer, pp 758–775

47. Goldwasser S, Rothblum GN (2012) How to compute in the presence of leakage. In: 53rd annual IEEE symposium on foundations of computer science, FOCS 2012, New Brunswick, NJ, USA, October 20–23, 2012, pp 31–40

48. Dziembowski S, Faust S (2011) Leakage-resilient cryptography from the inner-product extractor. In: Lee DH, Wang X (eds) Advances in cryptology - ASIACRYPT 2011 - 17th international conference on the theory and application of cryptology and information security, Seoul, South Korea, December 4–8, 2011. Proceedings. Lecture notes in computer science, vol 7073. Springer, pp 702–721

49. Balasch J, Faust S, Gierlichs B (2015) Inner product masking revisited. In: Oswald E, Fischlin M (eds) Advances in cryptology - EUROCRYPT 2015 - 34th annual international conference on the theory and applications of cryptographic techniques, Sofia, Bulgaria, April 26–30, 2015, Proceedings, Part I. Lecture notes in computer science, vol 9056. Springer, pp 486–510

50. Balasch J, Faust S, Gierlichs B, Paglialonga C, Standaert F-X (2017) Consolidating inner product masking. In: Takagi T, Peyrin T (eds) Advances in cryptology - ASIACRYPT 2017 - 23rd international conference on the theory and applications of cryptology and information security, Hong Kong, China, December 3–7, 2017, Proceedings, Part I. Lecture notes in computer science, vol 10624. Springer, pp 724–754

51. Cheng W, Carlet C, Goli K, Guilley S, Danger J-L (2019) Detecting faults in inner product masking scheme - IPM-FD: IPM with fault detection. Cryptology ePrint Archive, Report 2019/919. https://eprint.iacr.org/2019/919

52. Bringer J, Carlet C, Chabanne H, Guilley S, Maghrebi H (2014) Orthogonal direct sum masking – a smartcard friendly computation paradigm in a code, with builtin protection against side-channel and fault attacks. In: WISTP, Heraklion, Greece. LNCS, vol 8501. Springer, pp 40–56

53. Carlet C, Guilley S (2016) Complementary dual codes for counter-measures to side-channel attacks. Adv Math Commun 10(1):131–150 February
54. Poussier R, Guo Q, Standaert F-X, Carlet C, Guilley S (2017) Connecting and improving direct sum masking and inner product masking. In: Eisenbarth T, Teglia Y (eds) Smart card research and advanced applications - 16th international conference, CARDIS 2017, Lugano, Switzerland, November 13–15, 2017, Revised selected papers. Lecture notes in computer science, vol 10728. Springer, pp 123–141
55. Cedric T, Carlet C, Guilley S, Daif A (2018) Polynomial direct sum masking to protect against both SCA and FIA. J Cryptogr Eng
56. Goudarzi D, Rivain M (2017) How fast can higher-order masking be in software? In: Coron J-S, Nielsen JB (eds) Advances in cryptology - EUROCRYPT 2017 - 36th annual international conference on the theory and applications of cryptographic techniques, Paris, France, April 30 - May 4, 2017, Proceedings, Part I. Lecture notes in computer science, vol 10210, pp 567–597
57. Grosso V, Prouff E, Standaert F-X (2014) Efficient masked S-boxes processing - a step forward -. In: Pointcheval D, Vergnaud D (eds) Progress in cryptology - AFRICACRYPT 2014 - 7th international conference on cryptology in Africa, Marrakesh, Morocco, May 28–30, 2014. Proceedings. Lecture notes in computer science, vol 8469. Springer, pp 251–266
58. Ishai Y, Kushilevitz E, Li X, Ostrovsky R, Prabhakaran M, Sahai A, Zuckerman DI (2013) Robust pseudorandom generators. In: Automata, languages, and programming - 40th international colloquium, ICALP 2013, Proceedings, number PART 1 in lecture notes in computer science (including subseries lecture notes in artificial intelligence and lecture notes in bioinformatics), pp 576–588
59. Coron J-S, Greuet A, Zeitoun R (2019) Side-channel masking with pseudo-random generator. Cryptology ePrint Archive, Report 2019/1106. https://eprint.iacr.org/2019/1106
60. Herbst C, Oswald E, Mangard S (2006) An AES smart card implementation resistant to power analysis attacks, pp 239–252
61. Tillich S, Herbst C (2008) Attacking state-of-the-art software countermeasures—a case study for AES, pp 228–243
62. Benadjila R, Khati L, Prouff E, Thillard A (2019) Hardened library for AES-128 encryption/decryption on ARM Cortex M4 architecture. https://github.com/ANSSI-FR/SecAESSTM32

Part II
Spectral Approach in Side-Channel Attacks

In Part II of the book, we introduce a novel concept called *Spectral Approach* in the SCA topic. The contents are split into two chapters. Chapter 5 provides some definitions of the well-known spectral approach of pseudo-Boolean function computation. Chapter 6 provides a significant speed up of the CPA running time, by leveraging this approach.

Spectral Approach to Speed up the Processing

5

5.1 Walsh-Hadamard Transformation to Speed Up Convolution Computation

We provide hereafter the definition of the Walsh-Hadamard transform.

Definition 1 (*Walsh-Hadamard Transform* [1]) Let $f : \mathbb{F}_2^n \to \mathbb{R}$. The Walsh-Hadamard transform of f is a function $\hat{f} : \mathbb{F}_2^n \to \mathbb{R}$ defined as $\hat{f}(u) = \sum_{z \in \mathbb{F}_2^n} (-1)^{u \cdot z} f(z)$, where " $\cdot$ " is the canonical scalar product over spacevector $\mathbb{F}_2^n$.

There exists a butterfly algorithm to compute WHT using only additions (and subtractions), and no multiplications.

Furthermore, we denote by WHT the Walsh-Hadamard transform, which is the Fourier transform over the group $(\mathbb{S}, \oplus)$. Up to an unimportant coefficient $\frac{1}{\sqrt{2^n}}$, an example of WHT processing for $n = 3$ is illustrated in Fig. 5.1. It is noteworthy that WHT is its own inverse.

5.2 Convolution Product

Let us first recall the definition of a convolution product that can be evaluated quickly.

Definition 2 Let $(\mathbb{S}, \oplus)$ be a group of size 2^n. Let L, M be two functions from $\mathbb{S}$ to $\mathbb{R}$. A convolution product of L and M is the function denoted by $L \otimes M$ and defined from $\mathbb{S}$ to $\mathbb{R}$ by

$$\text{for every } k \in \mathbb{S}, \ (L \otimes M)(k) = \sum_{j \in \mathbb{S}} L(j) M(j \oplus k). \tag{5.1}$$

© The Author(s), under exclusive license to Springer Nature Switzerland AG 2021
M. Ouladj and S. Guilley, *Side-Channel Analysis of Embedded Systems*,
https://doi.org/10.1007/978-3-030-77222-2_5

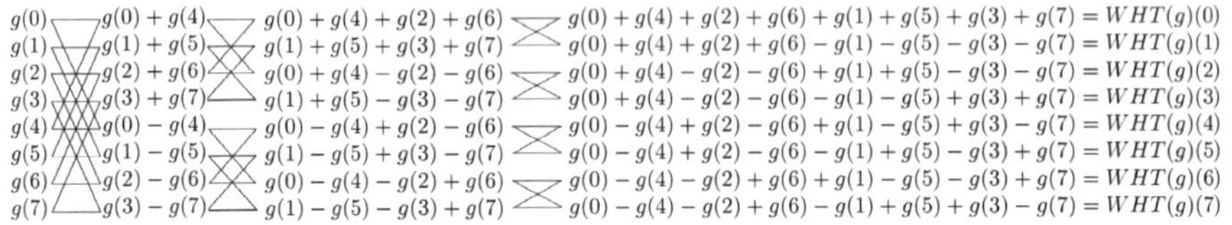

$$
\begin{array}{l}
g(0) \longrightarrow g(0)+g(4) \longrightarrow g(0)+g(4)+g(2)+g(6) \longrightarrow g(0)+g(4)+g(2)+g(6)+g(1)+g(5)+g(3)+g(7)=WHT(g)(0) \\
g(1) \longrightarrow g(1)+g(5) \longrightarrow g(1)+g(5)+g(3)+g(7) \longrightarrow g(0)+g(4)+g(2)+g(6)-g(1)-g(5)-g(3)-g(7)=WHT(g)(1) \\
g(2) \longrightarrow g(2)+g(6) \longrightarrow g(0)+g(4)-g(2)-g(6) \longrightarrow g(0)+g(4)-g(2)-g(6)+g(1)+g(5)-g(3)-g(7)=WHT(g)(2) \\
g(3) \longrightarrow g(3)+g(7) \longrightarrow g(1)+g(5)-g(3)-g(7) \longrightarrow g(0)+g(4)-g(2)-g(6)-g(1)-g(5)+g(3)+g(7)=WHT(g)(3) \\
g(4) \longrightarrow g(0)-g(4) \longrightarrow g(0)-g(4)+g(2)-g(6) \longrightarrow g(0)-g(4)+g(2)-g(6)+g(1)-g(5)+g(3)-g(7)=WHT(g)(4) \\
g(5) \longrightarrow g(1)-g(5) \longrightarrow g(1)-g(5)+g(3)-g(7) \longrightarrow g(0)-g(4)+g(2)-g(6)-g(1)+g(5)-g(3)+g(7)=WHT(g)(5) \\
g(6) \longrightarrow g(2)-g(6) \longrightarrow g(0)-g(4)-g(2)+g(6) \longrightarrow g(0)-g(4)-g(2)+g(6)+g(1)-g(5)-g(3)+g(7)=WHT(g)(6) \\
g(7) \longrightarrow g(3)-g(7) \longrightarrow g(1)-g(5)-g(3)+g(7) \longrightarrow g(0)-g(4)-g(2)+g(6)-g(1)+g(5)+g(3)-g(7)=WHT(g)(7)
\end{array}
$$

Fig. 5.1 Flow graph for computing WHT for $n = 3$ (no normalization coefficient)

The naïve computation of the convolution product for all possible keys k requires $\mathscr{O}(2^{2n})$ multiplications (one factor $\mathscr{O}(2^n)$ for the sum in Eq. (5.1), and another factor 2^n for all the keys in $\mathbb{S}$). Let us also denote the so-called *functions product* by $L \bullet M$ where $L \bullet M(j) = L(j)M(j), \forall j$. It is the coordinate-wise product.

An important property of the convolution product with respect to the Walsh-Hadamard transform is that, for every $k \in \mathbb{S}$, we happen to have

$$L \otimes M(k) = WHT^{-1}(WHT(L) \bullet WHT(M))(k). \tag{5.2}$$

The advantage of this property is that one can evaluate the convolution product for all possible values k at once, with an overall complexity of $\mathscr{O}(n2^n)$ additions instead of $\mathscr{O}(2^{2n})$ multiplication. Indeed, there is the conjunction of the fact that

- the WHT can be executed with depth n on 2^n inputs (as illustrated in Fig. 5.1) thanks to the butterfly structure, and that
- the WHT consists only in additions (or subtractions, which is the same in $\mathbb{F}_2^n$).

5.3 Extension of the Spectral Approach to the Higher Order

As will be shown, instead of computing the distinguisher for each subkey k (or equivalently for each sensitive value z) distinctly, only one computation is sufficient, taking advantage of the properties of the convolution product over a finite group. Let us first recall the definition of a convolution product that can be processed efficiently.

Definition 3 ([1]) Let $(\mathbb{S}, +)$ be a group of size 2^n. Let d be a non-zero integer. Let $f^{(0)}, \ldots f^{(d)}$ be $d + 1$ functions from $\mathbb{S}$ to $\mathbb{R}$. A dth-order convolution product of $f^{(0)}, \ldots, f^{(d)}$ is the function denoted by $f^{(0)} \otimes \cdots \otimes f^{(d)}$ and defined from $\mathbb{S}$ to $\mathbb{R}$ by: for every $z \in \mathbb{S}$,

$$(f^{(0)} \otimes \cdots \otimes f^{(d)})(z) \doteq \sum_{j_1 \in \mathbb{S}} \cdots \sum_{j_d \in \mathbb{S}} f^{(1)}(j_1) \cdots f^{(d)}(j_d) f^{(0)}(z - \sum_{w=1}^{d} j_w),$$

where $-z$ denotes the symmetric element of z relatively to the $+$ law (i.e., in the group $(\mathbb{S}, +)$).

The naïve computation of the convolution product for all possible z needs $\mathcal{O}(2^{dn})$ multiplications. In a view to speed up the processing of such convolution, let us also denote the so-called *functions product* by $f^{(1)} \bullet f^{(2)}$ where $f^{(1)} \bullet f^{(2)}(z) \doteq f^{(1)}(z) f^{(2)}(z)$. It is the associative coordinate-wise product. In addition, we denote by FFT the (fast) Fourier transform over the group $(\mathbb{S}, +)$. An important property of the convolution product with respect to the FFT is that

Proposition 1 (Convolution Theorem [1]) *For every $z \in \mathbb{S}$, we have*

$$(f^{(0)} \otimes \cdots \otimes f^{(d)})(z) = \frac{1}{2^n} FFT^{-1}\left(FFT(f^{(0)}) \bullet \cdots \bullet FFT(f^{(d)}) \right)(z). \quad (5.3)$$

Well-known result, provided hereafter for the chapter to be self-contained.

$$
\begin{aligned}
FFT^{-1}&\left(FFT(f^{(0)}) \bullet \cdots \bullet FFT(f^{(d)}) \right)(z) \\
&= \sum_u (-1)^{u \cdot z} FFT(f^{(1)})(u) \cdots FFT(f^{(n)})(u) \\
&= \sum_u (-1)^{u \cdot z} \sum_{y_1} (-1)^{y_1 \cdot u} f^{(1)}(y_1) \cdots \sum_{y_n} (-1)^{y_n \cdot u} f^{(n)}(y_n) \\
&= \sum_{y_1, \ldots, y_n} f^{(1)}(y_1) \cdots f^{(n)}(y_n) \sum_u (-1)^{u \cdot (z + y_1 + \cdots + y_n)} \\
&= \sum_{y_1, \ldots, y_n} f^{(1)}(y_1) \cdots f^{(n)}(y_n) \times 2^n \mathbb{1}_{y_1 + \cdots + y_n = z} \\
&= 2^n \sum_{\substack{y_1, \ldots, y_n \\ y_1 + \cdots + y_n = z}} f^{(1)}(y_1) \cdots f^{(n)}(y_n) \\
&= 2^n (f^{(0)} \otimes \cdots \otimes f^{(d)})(z).
\end{aligned}
$$

$\square$

The advantage of this property is that one can process the dth-order convolution product for all possible values z at once, with an overall complexity of $\mathcal{O}(dn2^n)$ *additions* instead of $\mathcal{O}(2^{dn})$ *multiplications*. Indeed, the computation of the complete spectrum benefits from the fast Fourier transform implemented by a butterfly algorithm.

The general expression of Eq. (5.3) can be formulated in terms of Walsh-Hadamard transform when $\mathbb{S} = \mathbb{F}_2^n$.

The two groups of interest for SCA's topic are $(\mathbb{F}_2^n, \oplus)$ for which the FFT coincides with the Walsh-Hadamard transform, and $(\mathbb{F}_{2^n}, +)$ for which it is the cyclical Fourier transform.

5.4 Conclusion

The spectral approach is power in terms of computational acceleration. In the following chapter, the incremental computation of the CPA will be speeded up, thanks to this spectral approach.

Reference

1. Terras A (1999) Fourier analysis on finite groups and applications. London mathematical society student texts. Cambridge University Press, Cambridge

Generalized Spectral Approach to Speed up the Correlation Power Analysis

6.1 Introduction

As shown previously in Sect. 3.2.3, the Correlation Power Analysis (CPA) is a method that allows to recover the secret information concealed in embedded devices [1]. It consists in leveraging the Pearson correlation coefficient as a way to relate an assumed model with the measured power consumed during the running of the operations that involve a secret subkey. Power analysis attacks are based on the principle that the instantaneous power consumption of a cryptographic device depends on the processed data and on the performed operations [2]. During a symmetric protocol, those operations are in particular processed by a non-linear function called S-box which is parametrized by a secret key and the involved message [3,4].

A method that relies on spectral analysis is presented in [5]. Its interest lies in the reduction of the attack computational complexity. Namely, if the key size equals n (typical the bitwidth of the cryptographic device), N is the required number of messages and D is the number of time samples per measure, then according to [6–8] the overall complexity of the CPA is about $\mathcal{O}(2^n N D)$. It is noticeable that the required number of messages N is very high in presence of countermeasures (high noise, random shuffling, random masking, ...). In such cases, values around $N = 10^8$ are not uncommon [6,9,10]. Furthermore for each trace a large number D of samples per trace is also required. When dealt with in a straightforward manner, this causes space memory and execution time problems.

Thanks to the spectral approach introduced by Guillot et al., if the adversary uses exactly one trace per possible message value (that is, there are $N = 2^n$ unique messages), then the complexity decreases from $\mathcal{O}(2^{2n} D)$ to $\mathcal{O}(n 2^n D)$. The disadvantage of this approach is the specific required set of measurements. Still, as we shall see, such approach can be generalized to any set of plaintexts, thanks to the coalescence principle (developed comprehensively in Part III). In the current chapter, we provide a generalization of this approach to any set of measurements. Our approach leads to a linear overall complexity $\mathcal{O}(N D)$ instead of $\mathcal{O}(2^n N D)$, when $N \gg n 2^n$.

© The Author(s), under exclusive license to Springer Nature Switzerland AG 2021
M. Ouladj and S. Guilley, *Side-Channel Analysis of Embedded Systems*,
https://doi.org/10.1007/978-3-030-77222-2_6

Furthermore, our method is extendible to already completed computations using incremental CPA versions (to avoid repetitive computations). It is also extendible to the higher order CPA (over masked implementations). Moreover the improvement is straightforwardly applicable to the covariance-based distinguisher.

6.1.1   Outline

The rest of the chapter is organized as follows. In Sect. 6.2, some preliminaries of CPA are recalled. In Sect. 6.3, the principle of the attack speed up is presented. This improvement is extended to the masking-based protected implementations in Sect. 6.4. These results are experimented on the AES S-box in Sect. 6.5. Finally, we come to a conclusion, in Sect. 6.6.

6.2   CPA's Preliminaries

For consistency reasons, the same notations as that of the previous chapters will be considered in the rest of this chapter.

6.2.1   Target of the Attack

Many symmetric ciphering algorithms such as DES [3] or AES [4] for instance are based on alternation between linear and non-linear computations. The non-linear stage is most often implemented in the form of S-boxes. In cryptographic algorithms, the S-boxes are actually a function $f : \{0, 1\}^n \times \{0, 1\}^n \rightarrow \{0, 1\}^{n'}$, whose inputs are a known data $x \in \{0, 1\}^n$ and an unknown secret subkey $k^\star \in \{0, 1\}^n$, and whose output is a value $z \in \{0, 1\}^{n'}$, namely,

$$z = f(x, k^\star).$$

With such setup, the attack consists in performing ciphering operations with various inputs x known by the adversary, then measuring the power consumption during the calculation and finally, trying to infer the values of the secret subkey $k^\star$. The adversary is assumed to know the device leakage model denoted $M(x, k)$ (for example, the Hamming weight model [1]). Let us notice that, very often, Hamming weight is assumed, but this is not restrictive because it is easy to adapt the suggested attack to other models. Thus, for a given calculus of an S-box with input x and secret subkey $k^\star$, the power consumption is related to the quantity $L(x)$ which is modeled by

$$L(x) = M(x, k^\star) + C + n_x, \tag{6.1}$$

where

- The first term $M(x, k^\star)$ is the model leakage, which can typically be the Hamming weight of $f(x, k^\star)$.

- The second term is a constant C that corresponds to the static average power consumption of the system and does neither depend on x nor on $k^\star$.
- The third term denoted n_x is a random noise due both to the measurement precision and to the difference between the leakage model and the real physical consumption; it is also independent from x and $k^\star$.

For the set of N messages, on which the measurements are performed, and for any candidate subkey k, the Pearson correlation coefficient [11] between the power consumption measurement and the consumption given by the leakage model M_k is defined as in [1] by

$$\rho(\mathbf{M_k}, \mathbf{L}) = \frac{\sum\limits_{i=1}^{N} \left(M(x_i, k) - \overline{M}_k\right)\left(L(x_i) - \overline{L}\right)}{\sqrt{\sum\limits_{i=1}^{N} \left(M(x_i, k) - \overline{M}_k\right)^2}\sqrt{\sum\limits_{i=1}^{N} \left(L(x_i) - \overline{L}\right)^2}} \in [-1, +1], \qquad (6.2)$$

such that $\overline{M}_k$ and $\overline{L}$ denote respectively the empirical means of M (for a given key k) and of L.

The CPA rationale is that, using a *good* leakage model L, the candidate subkey which leads to the highest correlation is the most probable one [9].

Let us consider that the adversary has the perfect leakage model (the worst situation for embedded system designers) with additive Gaussian noise. In this situation both CPA and covariance are the optimal distinguishers [12, Theorem 5].

In what follows, the computation complexity of ρ is studied and improved, in the context of SCA.

6.3 Carrying Out the CPA, with Arbitrary Set of Messages, According to the Spectral Approach

According to [6] the Pearson coefficient defined in Eq. (6.2) can be exactly rewritten as

$$\rho(\mathbf{M_k}, \mathbf{L}) = \frac{N s_5 - s_1 s_3}{\sqrt{(N s_2 - s_1^2)(N s_4 - s_3^2)}}, \qquad (6.3)$$

where $s_1 = \sum\limits_{i=1}^{N} L(x_i),$ $\qquad\qquad s_2 = \sum\limits_{i=1}^{N} L(x_i)^2,$

$s_3 = \sum\limits_{i=1}^{N} M(x_i, k),$ $\qquad\qquad s_4 = \sum\limits_{i=1}^{N} M(x_i, k)^2,$

and $s_5 = \sum\limits_{i=1}^{N} M(x_i, k)L(x_i).$

The computations of these sums can be carried out with an overall complexity of $\mathcal{O}(N2^n D)$ as stated recently in [6]. In order to further reduce this complexity one can benefit from an important property that holds true for the most key mixing-based cryptographic functions [13]. Namely, it is the Equal Images under different Subkeys (EIS).

Definition 1 (*Equal Images under different Subkeys (EIS)* [13, Def. 2]) Let V denotes an arbitrary set and let $G\colon\{0, 1\}^n \times \{0, 1\}^{n'} \to V$ be a function, such that for every subkey k all images $G(\{0, 1\}^n \times \{k\}) \subseteq V$ are equal. We say that the function F has the "Equal Images under different Subkeys (EIS)" property, if a function $\widetilde{F}\colon V \to \mathbb{R}$ exists such that $F = \widetilde{F} \circ G$.

A specific case is when $F(x, k)$ depends only on $x \oplus k$ as in common block ciphers (*i.e.*, on ciphers as DES [3] and AES [4]). Other cases are the 2^n-modular addition $(\mathrm{mod}(2^n))$, and the $2^{2^m} + 1$-modular multiplication of non-zero numbers, such that $m = 0, 1, 2, 3, 4$ (recall that $(2^{2^m} + 1)_{m=0,1,2,3,4}$ are the first Fermat numbers).

As will be shown, instead of computing the Pearson coefficient for each subkey k distinctly, only one computation is sufficient, taking advantage of the convolution product.

To analyze the CPA computation complexity, let us assume that the leakage model depends only on $x_i \oplus k$ (the bitwise addition). This is the case of the most common symmetric ciphers, where the EIS property holds. Furthermore, it will be generalized to any key mixing-based cryptographic function, at the end of the chapter.

For simplicity, we denote $M(x_i, k)$ by $M(x_i \oplus k)$ in the sequel.

For $s_1 = \sum\limits_{i=1}^{N} L(x_i)$, and $s_2 = \sum\limits_{i=1}^{N} L(x_i)^2$, the overall complexities of the straight-forward computations are about $\mathcal{O}(ND)$, so it is linear. The interest of our approach is in the s_3, s_4 and s_5 computations. To show the improvement, let us define $n(j) \doteq \#\{x_i \in \mathbb{S}; \ x_i = j\}$, where $\#E$ denotes the cardinality of the set E. In fact, $n(j)$ is the number of measurements corresponding to the same message $x_i = j$. So,

$$s_3 = \sum_{i=1}^{N} M(x_i \oplus k) = \sum_{j=0}^{2^n - 1} n(j) \cdot M(j \oplus k)$$

$$= n(.) \otimes M(.)(k)$$

$$= WHT^{-1} \left(WHT(n(.)) \bullet WHT(M(.)) \right)(k).$$

So, instead of computing s_3 for all possible values of k (with an overall complexity of $\mathcal{O}(N2^n)$), one can compute it faster using the latter result. In fact, as shown above, one can **pre-process** the computation of $WHT(M(.))$ only once, then during the attack the adversary computes $n(.)$, and eventually evaluates WHT^{-1} $(WHT(n(.)) \bullet WHT(M(.)))$ with an overall complexity of $\mathcal{O}(n2^n + N)$ instead of $\mathcal{O}(N2^n)$. It is noticeable that the $n(.)$ computation has an overall complexity of $\mathcal{O}(N)$.

Similarly,

$$s_4 = \sum_{i=1}^{N} M(k \oplus x_i)^2 = \sum_{j=0}^{2^n-1} n(j) \cdot M(j \oplus k)^2$$

$$= n(.) \otimes M^2(.)(k)$$

$$= WHT^{-1}\left(WHT(n(.)) \bullet WHT(M^2(.))\right)(k).$$

Thus, instead of computing s_4 for all possible values of k (an overall complexity of $\mathcal{O}(N2^n)$), one can compute it faster using the former result. That can be carried out by **pre-processing** the computation of $WHT(M^2(.))$ only once, and during the attack the adversary computes $WHT^{-1}\left(WHT(n(.)) \bullet WHT(M^2(.))\right)$. That leads also to the decrease of the overall complexity from $\mathcal{O}(N2^n)$ to $\mathcal{O}(n2^n + N)$.

For s_5 computation, one can follow a similar way by gathering traces of each class j in an artificial trace $\tilde{L}(j)$ (which will be termed "coalesced trace" in Part III), such that

$$\tilde{L}(j) = \begin{cases} \sum_{x_i=j} L(x_i) & \text{if } n(j) \neq 0 \\ 0 & \text{otherwise.} \end{cases}$$

So, one has

$$s_5 = \sum_{i=1}^{N} M(x_i \oplus k) \cdot L(x_i)$$

$$= \sum_{j=0}^{2^n-1} M(j \oplus k) \cdot \tilde{L}(j)$$

$$= M(.) \otimes \tilde{L}(.)(k)$$

$$= WHT^{-1}\left(WHT(M(.)) \bullet WHT(\tilde{L}(.))\right)(k).$$

That leads to an overall complexity of $\mathcal{O}(n2^n D + ND)$ additions instead of $\mathcal{O}(N2^n D)$ multiplications for the s_5 computation. The advantage of gathering the measures is that if the required number of traces N is large, one has to consider all of them, which is difficult (even impossible with high noise variance and counter-measures), both in the required space memory and in the processing time.

To summarize, according to these results, on can **pre-process** $WHT(M(.)$ and $WHT(M^2(.)$ then during the attack he processes $(s_i)_{1 \leq i \leq 5}$ to compute ρ according to Eq. (6.3). It is noticeable that s_3 and s_4 must be processed only once during the whole attack (independent from the trace dimensionality D). In contrast, s_1, s_2, and s_5 must be computed for each time sample of the traces.

Finally we conclude that carrying out the CPA following our approach leads to the decrease of the overall complexity from $\mathcal{O}(N2^n D)$ (as in [6]) to $\mathcal{O}(n2^n D + ND)$. Recall that typically $n = 8$ (ex. for byte-oriented AES) and $N = 10^6$. That means in such a case ($N \gg n2^n$), the complexity is about D millions instead of $256D$ millions.

6.3.1 Incremental CPA Computation

If an adversary (or a security evaluator) carries out a CPA with N_1 traces' set $\mathbb{S}_1$ without revealing the secret key, he must add some other N_2 traces' set $\mathbb{S}_2$. Carrying out a CPA incrementally means computing the current Pearson coefficient according to all the possible subkey values, without redoing the first computations which correspond to the N_1 traces' set. The Pearson coefficient expression in this case is also Eq. (6.3), such that $\mathbb{S} = \mathbb{S}_1 \cup \mathbb{S}_2$ and $N = N_1 + N_2$:

$$s_1 = \sum_{i=1}^{N} L(x_i) = \sum_{i=1}^{N_1} L(x_i) + \sum_{i=1}^{N_2} L(x_{N_1+i}) \doteq s_1' + s_1'', \text{ and}$$

$$s_2 = \sum_{i=1}^{N} L(x_i)^2 = \sum_{i=1}^{N_1} L(x_i)^2 + \sum_{i=1}^{N_2} L(x_{N_1+i})^2 \doteq s_2' + s_2''. \text{ So, one can update } s_1$$

and s_2 with an overall complexity of $\mathcal{O}(N_2 D)$ without redoing the first part of the computation.

For s_3, s_4 and s_5 computations, let us define $n_1(j) \doteq \#\{x_i \in \mathbb{S}_1; \ x_i = j\}$ and $n_2(j) \doteq \#\{x_{N_1+i} \in \mathbb{S}_2; \ x_{N_1+i} = j\}$.

$$s_3 = \sum_{i=1}^{N} M(k \oplus x_i)$$

$$= \sum_{i=1}^{N_1} M(k \oplus x_i) + \sum_{i=1}^{N_2} M(k \oplus x_{N_1+i})$$

$$= \sum_{j=0}^{2^n-1} n_1(j) \cdot M(j \oplus k) + \sum_{j=0}^{2^n-1} n_2(j) \cdot M(j \oplus k)$$

$$= n_1(.) \otimes M(.)(k) + n_2(.) \otimes M(.)(k)$$

$$= WHT^{-1}\left(WHT(n_1) \bullet WHT(M(.))\right)(k) + WHT^{-1}\left(WHT(n_2) \bullet WHT(M(.))\right)(k)$$

$$\doteq s_3' + s_3''.$$

This result actually holds valid owing to the linearity of the Walsh-Hadamard transform. So, one can update s_3 with an overall complexity of $\mathcal{O}(n2^n + N_2)$ without redoing the first part of the computation.

Similarly,

$$s_4 = \sum_{i=1}^{N} M(k \oplus x_i)^2$$

$$= \sum_{i=1}^{N_1} M(k \oplus x_i)^2 + \sum_{i=1}^{N_2} M(k \oplus x_{N_1+i})^2$$

$$= \sum_{j=0}^{2^n-1} n_1(j) \cdot M(j \oplus k)^2 + \sum_{j=0}^{2^n-1} n_2(j) \cdot M(j \oplus k)^2$$

$$= n_1(.) \otimes M(.)^2(k) + n_2(.) \otimes M(.)^2(k)$$

$$= WHT^{-1}\left(WHT(n_1) \bullet WHT(M(.)^2)\right)(k) + WHT^{-1}\left(WHT(n_2) \bullet WHT(M(.)^2)\right)(k)$$

$$\doteq s_4' + s_4''.$$

So, one can also update s_4 with an overall complexity of $\mathcal{O}(n2^n + N_2)$ without redoing the first part of the computation.

For computing s_5 in incremental way, let us gather traces of the jth class into a single artificial trace $\tilde{L}_1(j)$ for $\mathbb{S}_1$ and the same thing for $\mathbb{S}_2$, such that

$$\tilde{L}_1(j) = \begin{cases} \sum_{x_i=j} L(x_i) & \text{if } n_1(j) \neq 0 \\ 0 & \text{otherwise} \end{cases},$$

$$\tilde{L}_2(j) = \begin{cases} \sum_{x_{N_1+i}=j} L(x_{N_1+i}) & \text{if } n_2(j) \neq 0 \\ 0 & \text{otherwise} \end{cases},$$

one has

$$s_5 = \sum_{i=1}^{N} M(x_i \oplus k) \cdot L(x_i)$$

$$= \sum_{i=1}^{N_1} M(x_i \oplus k) \cdot L(x_i) + \sum_{i=1}^{N_2} M(x_{N_1+i} \oplus k) \cdot L(x_{N_1+i})$$

$$= \sum_{j=0}^{2^n-1} M(j \oplus k) \cdot \tilde{L}_1(j) \sum_{j=0}^{2^n-1} M(j \oplus k) \cdot \tilde{L}_2(j)$$

$$= M(.) \otimes \tilde{L}_1(.)(k) + M(.) \otimes \tilde{L}_2(.)(k)$$

$$= WHT^{-1}\left(WHT(M(.)) \bullet WHT(\tilde{L}_1)\right)(k) + WHT^{-1}\left(WHT(M(.)) \bullet WHT(\tilde{L}_2)\right)(k)$$

$$\doteq s_5' + s_5''.$$

So, one can update s_5 with an overall complexity of $\mathcal{O}(n2^n D + N_2 D)$ without redoing the first part of the computation.

That leads us to conclude: following our approach, one can update the ρ value with an overall complexity of $\mathcal{O}(n2^n D + N_2 D)$ instead of $\mathcal{O}(N_2 2^n D)$.

Same Improvements for the covariance-based distinguisher: The covariance-based distinguisher consists in nothing but computing the numerator for the Pearson coefficient:

$$COV(\mathbf{M_k}, \mathbf{L}) = \frac{1}{N} \sum_{i=1}^{N} (M(x_i, k) - \overline{M}_k)(L(x_i) - \overline{L}) = N s_5 - s_1 s_3. \tag{6.4}$$

Thereby, our improvements hold also for this distinguisher by improving the computations of s_1, s_3, and s_5.

6.4   Extension of the Improvements to the Protected Implementations by Masking

Let us consider a leaking device that runs a cryptographic algorithm protected by *masking* [14]. Let d denotes a strictly positive integer. The protection by a masking of order d consists in dividing each sensitive variable Z into $d + 1$ shares $Z^{(0)}, \ldots, Z^{(d)}$. In practice at least two different kinds of masking can be used: the *Boolean* and the *arithmetic* masking schemes [15]. Let us denote by $L^{(w)}$ the measured leakage sample corresponding to the w-th share of the sensitive variable Z, such that $0 \leq w \leq d$. In order to recover information on Z when a d-th order masking is applied, an adversary must combine the leakages from all the $d + 1$ shares in a so-called $(d + 1)$-th high-order SCA.

Now, in order to conduct a higher order attack, many combination functions (which aim at combining the leakages of the different shares to form a new exploitable signal) were introduced in the literature (refer to Sect. 3.3.1). According to [16], the optimal combination technique against the first-order Boolean masking is the *normalized* product $\left((L^{(0)} - \mathbb{E}(L^{(0)}))(L^{(1)} - \mathbb{E}(L^{(1)})) \right)$. The authors of [16] show that this combination function should be accompanied by $\mathbb{E}\left((M_k^{(0)} - \mathbb{E}(M_k^{(0)}))(M_k^{(1)} - \mathbb{E}(M_k^{(1)})) \mid Z \right)$ as an optimal model. A Higher Order Optimal Distinguisher (HOOD) is introduced in [17]. It states that the CPA with *normalized* product combination is optimal for high noise, independently from the masking technique and the masking order.

While the attack is carried out on the combination leakage, as suggested in [16, 17], and explained hereabove, all our ideas developed in the previous section to improve the CPA computation efficiency can straightforwardly be applied against these protected devices.

6.5   Experiments

In order to assess our improvement we compare the running time of CPA using our approach by the running time of the approach described by the 2017 JCEN paper [6]. The target value of our experience is a first round S-box output of a simulated running AES. In practice the adversary does not precisely know the time sample when the target value is computed. So, she must compute the correlation coefficient within some time window. In our experiences we computed it in a window made up of $D = 1000$ time samples. As shown in Fig. 6.1 the Running time improvement tends toward 100 when the number of measurements is sufficiently large ($N \gg n2^n$). This ratio is due to the overall complexity of the computation of s_1, s_2, and s_5 using our improvements (respectively, $ND + ND + ND = 3ND$) and the computation of the same sums without using our improvements (respectively, $ND + ND + 256ND = 258ND$). That leads to the ratio $258/3$ which is not far from the resulting value of

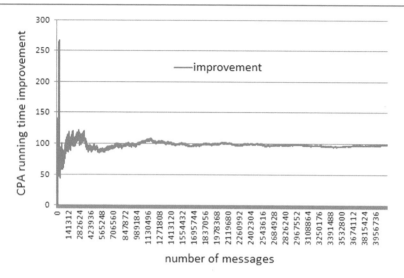

Fig. 6.1 CPA running time improvement (ratio), according to the number of measurements

100. Recall that s_3 and s_4 computations are independent from D and they do not affect the overall complexity.

Similar experiment has been done for the covariance-based distinguisher. We observed a similar improvement ratio. Figure 6.2 shows that the running time improvement tends toward 106 when the number of measurements is sufficiently large. This ratio is due to the overall complexity of the computation of s_1 and s_5 using our improvements (respectively, $ND + ND = 2ND$) and the computation of the same sums without using our improvements (respectively, $ND + 256ND = 257ND$). That leads to the ratio $257/2$ which is not far from the resulting value of 106. Recall that s_3 computation is independent from D and it does not affect the overall complexity.

6.6 Conclusion

In this chapter a new running time complexity is provided for both CPA and covariance distinguishers. To the best of our knowledge the CPA/covariance complexity that we provide in this chapter is the best known one. Since the new complexity is linear in the size of data (the size of measurements equals ND), one can conclude that no significant improvement will still be possible because the adversary needs at least to read all the measurements.

It is noteworthy that all our results that assume the group operation $\oplus$ over the set $\{0, 1\}^n$ (block cipher) holds true for any other group operation $\odot$ as long as Walsh-Hadamard Transform is replaced by Fourier Transform on this group ($\{0, 1\}^n, \odot$). For

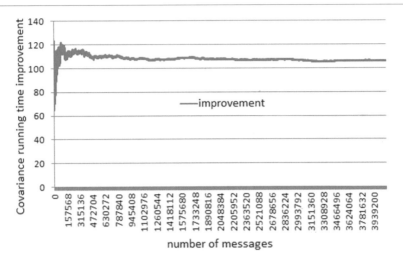

Fig. 6.2 Covariance running time improvement (ratio), according to the number of measurements

example, if the operation is $+$ (mod2^n), Cyclical Fourier Transform (FFT) should replace Walsh-Hadamard Transform (WHT).

In the next part of this book, we will show how and why the adversary could use the EIS property for gathering the traces corresponding to the same message in one averaged trace [13]. This gathering leads to a significant reduction both of the memory space and the running time complexities. Furthermore, it allows the adversary to compute several distinguishers according to the spectral approach, which leads to an even more significant improvement in terms of running time complexity.

References

1. Brier É, Clavier C, Olivier F (2004) Correlation power analysis with a leakage model. In: Joye M, Quisquater J-J (eds), Cryptographic hardware and embedded systems - CHES 2004: 6th international workshop Cambridge, MA, USA, August 11–13, 2004. Proceedings. Lecture notes in computer science, vol 3156. Springer, pp 16–29
2. Messerges TS, Dabbish EA, Sloan RH (2002) Examining smart-card security under the threat of power analysis attacks. IEEE Trans Comput 51(5):541–552
3. NIST/ITL/CSD. Data encryption standard. FIPS PUB 46-3, Oct 1999. http://csrc.nist.gov/publications/fips/fips46-3/fips46-3.pdf
4. NIST/ITL/CSD. Advanced encryption standard (AES). FIPS PUB 197, Nov 2001. http://nvlpubs.nist.gov/nistpubs/FIPS/NIST.FIPS.197.pdf (also ISO/IEC 18033-3:2010)
5. Guillot P, Millérioux G, Dravie B, El Mrabet N, Spectral approach for correlation power analysis. In: Hajji et al. [100], pp 238–253
6. Bottinelli P, Bos JW (2017) Computational aspects of correlation power analysis. J Cryptograph Eng 7(3):167–181

7. Schimmel O, Duplys P, Boehl E, Hayek J, Rosenstiel W (2010) Correlation power analysis in frequency domain. In: COSADE, pp 1–3, February 4–5 2010. Darmstadt, Germany. http://cosade2010.cased.de/files/proceedings/cosade2010_paper_1.pdf

8. Emmanuel P (ed) (2013) Constructive side-channel analysis and secure design - 4th international workshop, COSADE 2013, Paris, France, March 6–8, 2013, Revised selected papers. Lecture notes in computer science, vol 7864. Springer

9. Bilgin B, Gierlichs B, Nikova S, Nikov V (2014) Rijmen V. Higher-order threshold implementations 8874:326–343

10. Moradi A, Poschmann A, Ling S, Paar C (2011) Wang H. Pushing the limits: a very compact and a threshold implementation of aes 6632:69–88

11. Pearson K (1900) On lines and planes of closest fit to points in space. Philos Mag 2:559–572

12. Heuser A, Rioul O, Guilley S, Good is not good enough - deriving optimal distinguishers from communication theory. In: Batina and Robshaw [14], pp 55–74

13. Werner S, Kerstin L, Paar AC (2005) Model stochastic, for differential side channel cryptanalysis. In: LNCS (ed), CHES. LNCS, vol 3659 pp. 30–46. Springer, Edinburgh

14. Piret G, Standaert F-X (2008) Security analysis of higher-order Boolean masking schemes for block ciphers (with conditions of perfect masking). IET Inf Secur 2(1):1–11. https://doi.org/10.1049/iet-ifs:20070066

15. Kerstin L, Kai S, Paar C (2004) DPA, on n-bit sized Boolean and arithmetic operations and its application to IDEA, RC6, and the HMAC-construction. In: CHES. Lecture notes in computer science, vol 3156. Springer, Cambridge, pp. 205–219

16. Prouff E, Rivain M, Bevan R (2009) Statistical analysis of second order differential power analysis. IEEE Trans Comput 58(6):799–811

17. Bruneau N, Guilley S, Heuser A, Rioul O (2014) Masks will fall off – higher-order optimal distinguishers. In: Sarkar P, Iwata T, (eds), Advances in cryptology – ASIACRYPT 2014 - 20th international conference on the theory and application of cryptology and information security, Kaoshiung, Taiwan, R.O.C., December 7–11, 2014, Proceedings, Part II. Lecture notes in computer science, vol 8874. Springer, pp 344–365

Part III
Coalescence-based Side-Channel Attacks

Part III is organized in three chapters. In Chap. 7, we introduce the so-called *coalescence principle* and its advantages, especially we prove the asymptotic equivalence between SCAs with and without respect to this principle. Then, we explain its compatibility with the *spectral approach* introduced in [98], by studying the improvement of the CPA and the covariance-based distinguisher.

In Chap. 8, we introduce several significant improvements of the LRA implementation. Some of these improvements are obtained thanks to this duo "coalescence principle and spectral approach". These improvements hold both in profiled and non-profiled LRA, either in first or higher order. Furthermore, we exhibit the relationship between the LRA and the Numerical Normal Form (NNF) [43], which has been originally introduced in the field of Boolean functions. Leveraging this relation, we show that it is possible to exhibit the polynomial expression of combination function against the first-order additive masking. In fact, knowing such a polynomial expression is useful for the H-O CPA. Also, knowing the degree of the polynomial is useful for the adversary to choose the optimal basis to make the best out of the LRA.

Recall that the coalescence principle is introduced by Schindler *et al.* in the first LRA paper since 2005 [187], but without proof of the approach. Moreover, it has not been extended to other distinguishers. In Chap. 9, we prove for the first time, how the template attack can be formulated according to the coalescence principle. Consequently, TA benefits significantly from the advantage of the winning duo "coalescence principle and spectral approach". Furthermore, we introduce a new approach to extend the template attacks to the masking-based protected implementations. This extension is also compatible with all the introduced improvements.

Coalescence Principle

7

Notations Throughout this chapter we use the same notations as above (see Sect. 2.1). Recall that, during an attack, we consider that the adversary targets the manipulation of a single sensitive variable Z, such that $Z = F(X, k)$. Typically $Z = sbox(X \oplus k)$, such that $s - box$ denotes a substitution box and $\oplus$ denotes the bitwise addition. The attack is carried out with N traces $\vec{l}_0, \ldots, \vec{l}_{N-1}$. Each $\vec{l}_i \hookleftarrow \vec{L}$ corresponds to the processing of $z_i = F(x_i, k)$. The number of samples per traces (instantaneous leakage points) is denoted by D.

7.1 Difference Between SCAs with and Without Coalescence

Let us first study the difference between the distribution of raw traces and that of the averaged ones per class (a so-called coalescence approach [1]). From Cochran's Theorem [2] one has the following proposition:

Proposition 7.1 (Simplified version of Cochran's Theorem) *Let* $(l_i)_{0 \leq i \leq N-1}$ *be* N *univariate observations of a random variable* L *that follows a normal distribution* $\mathcal{N}(\mu, \sigma)$. *Let* $\hat{\mu}$ *and* $\hat{\sigma}^2$ *be respectively the estimators of* μ *and* σ^2 *defined by* $\hat{\mu} = \frac{1}{N} \sum_{i=0}^{N-1} l_i$, *and* $\hat{\sigma}^2 = \frac{1}{N} \sum_{i=0}^{N-1} (l_i - \mathbb{E}(L))^2$. *Then, one has*

- $\hat{\mu}$ *and* $\hat{\sigma}^2$ *are independents random variables,*
- $\hat{\mu} \sim \mathcal{N}(\mu, \sigma/\sqrt{N})$ *and* $\frac{n\hat{\sigma}}{\sigma^2} \sim \chi^2(N - 1)$.

An important result of this book is the following proposition.

Proposition 7.2 *Under the additive independent noise assumption, carrying out an SCA with or without coalescence is asymptotically equivalent.*

© The Author(s), under exclusive license to Springer Nature Switzerland AG 2021
M. Ouladj and S. Guilley, *Side-Channel Analysis of Embedded Systems*,
https://doi.org/10.1007/978-3-030-77222-2_7

To compare the efficiency of an SCA with and without coalescence, let us consider that, under the additive independent noise assumption, the adversary knows the leakage model. In this situation both CPA and covariance distinguishers are the optimal distinguishers [3]. To prove Proposition 7.2, we limit ourselves to these two optimal distinguishers.

Proof Under the additive independent noise assumption, the leakage $\vec{l}_i$ is divided into two independent parts, such that $\vec{l}_i = \mathbb{E}(\vec{L}|x_i) + b_i$, where $\mathbb{E}(\vec{L}|x_i)$ is the determinist part and b_i is the noise. Let us denote the leakage model used in the attack by $g_k(x_i)$ (that corresponds to the sensitive variable $Z = x_i \oplus k$, for example). So, the covariance distinguisher is

$$
\frac{1}{N}\sum_{i=0}^{N-1} g_k(x_i) \cdot \vec{l}_i = \frac{1}{N}\sum_{i=0}^{N-1} g_k(x_i) \cdot \left(\mathbb{E}(\vec{L}|x_i) + b_i \right)
$$

$$
= \frac{1}{N}\sum_{i=0}^{N-1} g_k(x_i) \cdot \mathbb{E}(\vec{L}|x_i) + \frac{1}{N}\sum_{i=0}^{N-1} g_k(x_i) \cdot b_i
$$

$$
= \frac{1}{N}\sum_{j=0}^{2^n-1} n_j \cdot g_k(j) \cdot \mathbb{E}(\vec{L}|j) + \frac{1}{N}\sum_{j=0}^{2^n-1} g_k(j) \cdot (\sum_{x_i=j} b_i).
$$

Such that $n_j = \#\{x_i;\ x_i = j\}$.

By regrouping b_i in the classes (j), one can note $\sum_{x_i=j} b_i$ by $\sum_{c=1}^{n_j} b_{j,c}$. So, the covariance distinguisher becomes

$$
\frac{1}{N}\sum_{i=0}^{N-1} g_k(x_i) \cdot \vec{l}_i = \frac{1}{N}\sum_{j=0}^{2^n-1} n_j \cdot g_k(j) \cdot \mathbb{E}(\vec{L}|j) + \frac{1}{N}\sum_{j=0}^{2^n-1} g_k(j) \cdot (\sum_{c=1}^{n_j} b_{j,c})
$$

$$
= \frac{1}{2^n}\sum_{j=0}^{2^n-1} \frac{n_j 2^n}{N} \cdot g_k(j) \cdot \left(\mathbb{E}(\vec{L}|j) + \frac{\sum_{c=1}^{n_j} b_{j,c}}{n_j} \right),
$$

such that the coalesced traces of the jth class is $\frac{1}{n_j}\sum_{x_i=j} \vec{l}_i = \mathbb{E}(\vec{L}|j) + \frac{\sum_{c=1}^{n_j} b_{j,c}}{n_j}$ if $n_j \neq 0$ (the coalesced traces equals 0 if $n_j = 0$). So, up to a weight factor per class $\left(\frac{n_j 2^n}{N} \right)$, the traces being coalesced or not, we have the same distinguisher with a different noise variance $\left(\frac{\sum_{c=1}^{n_j} b_{j,c}}{n_j} \text{ instead of } b_j \right)$.

In fact, studying the raw traces distribution means studying a random variable with a fixed standard deviation σ and an arbitrary number N of traces. If the required number of traces N is large, one has to consider all of them, which is difficult (even impossible in high noise variance and countermeasures). By the Law of Large Numbers (LLN), doing the coalescence, with equal probabilities of each possible message $p = \frac{1}{2^n}$, gives $\frac{n_j 2^n}{N}$ tending toward 1. Furthermore, in such a case the number of the artificial traces (grouped by class) is at most 2^n, and according to Cochran Theorem 7.1, the noise standard deviation becomes $\sqrt{\frac{\sigma}{n_j}}$. That leads to a possible (even a fast) attack processing. Finally we conclude that carrying out the attack with averaging traces means reducing the standard deviation of the artificial random variable (by averaging traces per j values) instead of increasing the number of raw traces.

The asymptotic equivalence between SCA with and without coalesced traces holds also for CPA. This comes as a consequence of the following important property of Pearson's correlation coefficient:

Proposition 7.3 *([4, Proposition 5]) Let $\vec{L}$ and X be two random variables. Then, for every function g_k defined over $\mathbb{Z}$, we have*

$$\rho(g_k(X), \vec{L}) = \rho\left(g_k(X), \mathbb{E}\left(\vec{L}|X\right)\right) \times \rho\left(\mathbb{E}\left(\vec{L}|X\right), \vec{L}\right). \tag{7.1}$$

So, an adversary that carrying out a CPA on leakage traces $\vec{l_i} \hookleftarrow \vec{L}$ with a leakage model function $g_k(x_i)$, such that $x_i \hookleftarrow X$, will estimate $\rho(g_k(X), \vec{L}) = \rho\left(g_k(X), \mathbb{E}\left(\vec{L}|X\right)\right) \times \rho\left(\mathbb{E}\left(\vec{L}|X\right), \vec{L}\right)$. Since $\rho\left(\mathbb{E}\left(\vec{L}|X\right), \vec{L}\right)$ is independent from k, she can carry out a CPA by only estimating $\rho\left(g_k(X), \mathbb{E}\left(\vec{L}|X\right)\right)$. This approach holds if and only if the adversary has a good estimator of $\mathbb{E}\left(\vec{L}|X\right)$. This estimator can be obtained by simply averaging the traces per x_i values. In other terms, the adversary can carry out a CPA over the coalesced measurements by processing $\rho\left(g_k(X), \mathbb{E}\left(\vec{L}|X\right)\right)$ instead of carrying it out with the raw traces by processing $\rho(g_k(X), \vec{L})$. So one can conclude the asymptotic equivalence between SCA with and without coalescence.   ◇   □

So, for carrying out a CPA (resp. Covariance-based SCA) one can compute the incremental ρ as in Eq. (6.3) (resp. the incremental COV as in Eq. (6.4)), where

$$s_1 = \sum_{i=1}^{2^n} L(x_i), \qquad\qquad s_2 = \sum_{i=1}^{2^n} L(x_i)^2,$$

$$s_3 = \sum_{i=1}^{2^n} M(i, k), \qquad\qquad s_4 = \sum_{i=1}^{2^n} M(i, k)^2,$$

$$\text{and } s_5 = \sum_{i=1}^{2^n} M(i, k)L(x_i).$$

This leads to a significant improvement in the running time, while $N >> 2^n$ (N becomes 2^n and $n(.)$ is an asymptotically constant function). Furthermore, if the sensitive variable depends only on $i \oplus k$, then s_3 and s_4 become asymptotically independent from the key k and s_5 can be written as a convolution $M(.) \otimes L(.)(k)$. That leads to an even more speed up of the attack.

Due to this asymptotic equivalence between SCAs with and without coalescence, LRA (introduced first with coalescence) also tends toward LRA without coalescence. This equivalence will be shown experimentally in Chap. 8 both for LRA and CPA.

Since the Cochran theorem holds for the Gaussian random vectors, the averaging traces by class principle can be applied in the multidimensional analysis, as will be shown in Chap. 9.

7.2 Optimization of the Stochastic Collision Attack Thanks to the Coalescence

In this section we review [5] with a coalescence point of view.

7.2.1 Preliminaries

In the sequel, we are interested in block ciphers which can be attacked by collisions. The access to substitution boxes (S-Boxes) is especially leaky, because they consist in a memory look-up, therefore favorable case studies are block ciphers which reuse the same instance of S-Box several times. This is the case for AES and PRESENT. We denote by n the fan-in (i.e., the number of inputs) of the S-Box, that is $n = 8$ bits for AES and $n = 4$ bits for PRESENT. The implementation can be either hardware or software (in which case the same memory is recalled sixteen times). For $w \in \{1, \ldots, W\}$ (e.g., $W = 16$ for AES), we denote by $k^{*(w)} \in \mathbb{F}_2^n$ the w-th secret key byte and by $k^{(w)}$ any possible key hypothesis thereof. The w-th byte coordinate of the plaintext corresponding to the qth query ($1 \le q \le N$) is denoted by $x_q^{(w)}$. The associated leakage is denoted by $l_q^{(w)}$.

For all the indices w, the leakage models are assumed to be identical[1] but unknown, which leads to view $l_q^{(w)}$ as realizations of the following random variable:

$$L^{(w)} = \phi(X^{(w)} \oplus k^{*(w)}) + B^{(w)} \ , \tag{7.2}$$

where the noise B is independent among w as well as q and $\phi = M \circ Z$ is a composition of two deterministic functions. In particular, the S-Box function $Z : \mathbb{F}_2^n \to \mathbb{F}_2^n$ is algorithmic specific, whereas $M : \mathbb{F}_2^n \to \mathbb{R}$ is an unknown function related to the device architecture.

In the rest of the paper, we will additionally employ the following shortcut notations:

---

[1] This assumption is reasonable when the code or the hardware is reused; this occurs in practice for the sake of cost overhead mitigation.

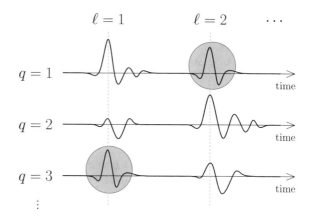

Fig. 7.1 Setup for a collision attack. The highlighted parts of the traces (blue circles) correspond to the following collision: $x_1^{(2)} \oplus k^{(2)} = x_3^{(1)} \oplus k^{(1)}$.

- $\vec{k}$ denotes the vector $(k^{(1)}, \cdots, k^{(W)})$,
- $\vec{l}^{(\cdot)}$ is the $N \times W$ matrix whose rows correspond to W-variate leakages,
- to each matrix element $l_q^{(w)}$ of $\vec{l}^{(\cdot)}$ is associated a plaintext element $x_q^{(w)}$ and key element $k^{(w)}$ (note that, as usual, for each w the key element $k^{(w)}$ is assumed to be the same for all the $x_q^{(w)}$ and $l_q^{(w)}$,
- $\vec{l}^{(w)}$ is a list of the N leakages which are the consequence of the processing of $(x_q^{(w)})_{1 \leq q \leq N}$,
- $\vec{l}_q$ is a list of the W leakages which are the consequence of the processing of $(x_q^{(w)})_{1 \leq w \leq W}$. It is the qth row of $\vec{l}^{(\cdot)}$. The corresponding vector of plaintext elements is denoted by $\vec{x}_q$.

For any vector $\vec{y} \in \mathbb{R}^W$, we will denote by $\|\vec{y}\|$ the *Euclidean norm* of $\vec{y}$. Moreover, we will denote by f_{μ,σ^2} the *Normal distribution* with mean μ and standard deviation σ. The notation will be simplified into f_{σ^2} when the mean is assumed to be null (note that for any y we have $f_{\mu,\sigma^2}(y) = f_{\sigma^2}(y - \mu)$). Eventually, for any function φ defined from a set E to a set F and any vector $\vec{y} \doteq (y_1, \cdots, y_W) \in E^W$, we will simply denote by $\varphi(\vec{y})$ the W-dimensional vector $(\varphi(y_1), \cdots, \varphi(y_W))$.

7.2.2 New Concept of Stochastic Collision Distinguisher

A *distinguisher* is a function which takes the known plaintexts and the measured leakages and returns a key guess. A specific distinguisher is *optimal* when maximizing the success probability of the attack [3]. Besides considering univariate first-order distinguisher, two other scenarios have been considered so far: first, in [6] the authors derived optimal distinguishers when applied on masking countermeasures and second, [7] studied optimal distinguisher when dimension reduction is helpful. Thus, both previous studies considered multiple leaking points that are related to only one key byte and plaintext.

In the next proposition we derive the optimal distinguisher in the scenario of collision attacks, namely considering W independent leakage samples with W different key hypotheses and plaintexts. For the paper to be self-content, we recall in Fig. 7.1 the principle of a collision in a side-channel context.

Proposition 7.4 (Optimal collision based distinguisher) *The optimal collision based distinguisher, when the noise is Gaussian and isotropic in the W samples, namely $B^{(\cdot)} \sim \mathcal{N}(0, \mathrm{diag}(\sigma^2, \ldots, \sigma^2))$, does not depend on σ^2, and can be expressed as*

$$\mathcal{D}_{opt}(\vec{x}^{(\cdot)}, \vec{l}^{(\cdot)}) = \underset{\vec{k} \in (\mathbb{F}_2^n)^W}{argmin} \sum_{q=1}^{N} \left\| \vec{l}_q - \phi(\vec{x}_q \oplus \vec{k}) \right\|^2 . \tag{7.3}$$

Proof The optimal distinguisher $\mathcal{D}_{opt}$ is the ML (Maximum Likelihood). We write p for random variables' densities. For example, p_B denotes the density of B. We can thus express the ML distinguisher as follows:

$$\mathcal{D}_{opt}(\vec{x}^{(\cdot)}, \vec{l}^{(\cdot)}) = \underset{\vec{k} \in (\mathbb{F}_2^n)^W}{argmix} \, p(\vec{l}^{(\cdot)} | \vec{x}^{(\cdot)}, \vec{k})$$

$$= \underset{\vec{k} \in (\mathbb{F}_2^n)^W}{argmix} \prod_{q=1}^{N} p_{B_q^{(\cdot)}} \left(\vec{l}_q - \phi(\vec{x}_q \oplus \vec{k}) \right) \tag{7.4}$$

$$= \underset{\vec{k} \in (\mathbb{F}_2^n)^W}{argmix} \sum_{q=1}^{N} \log \prod_{w=1}^{W} f_{\sigma^2}(l_q^{(w)} - \phi(x_q^{(w)} \oplus k^{(w)})) , \tag{7.5}$$

which implies (7.3) by simply developing the normal distribution law $f_{\sigma^2}(\cdot)$, and by removing the key-independent terms. Basically, Eq. (7.4) results from the noise independence from trace to trace. Eq. (7.5) arises from the Gaussian distribution of $B^{(\cdot)}$, which is assumed *isotropic*, and from the fact that the logarithm is a strictly increasing function. □

Remark 7.1 The hypothesis of noise isotropy in the sample space is often satisfied in practice. This can be justified as follows. Assuming that the thermal noise is white, the correlation window length is inversely proportional to the bandwidth of the measurement apparatus (typically 1 GHz). Therefore, the correlation between noisy samples does not extend beyond 1 ns. In practice, the S-box calls are at least separated by one clock period that is 10 to 100 ns. Therefore, the measurement noise samples are independent from one call to another. Still, if the W noise samples are correlated, Proposition 7.4 would still hold, by trading the Euclidean distance by the Mahalanobis distance [8].

Unfortunately, Eq. (7.3) has one major drawback, as discussed for other optimal distinguishers in [3,6,7] it can only be computed provided the leakage model (the function ϕ) is completely known. However, naturally, the strength of the state-of-the-art collision attacks is the unnecessariness of the knowledge of the leakage model.

To cope with this problem we extend Proposition 7.4 to the case where the model becomes part of the *secrets* to be guessed as similarly done in LRA.

Definition 7.1 *(Stochastic collision distinguisher)*

$$\mathscr{D}_{sto.coll}(\vec{x}^{(\cdot)}, \vec{l}^{(\cdot)}) = \underset{\vec{k} \in (\mathbb{F}_2^n)^W}{\operatorname{argmin}} \min_{\phi:\mathbb{F}_2^n \to \mathbb{R}} \sum_{q=1}^{N} \left\| \vec{l}_q - \phi(\vec{x}_q \oplus \vec{k}) \right\|^2 ,$$

where ϕ lives in the vector space of the functions $\mathbb{F}_2^n \to \mathbb{R}$.

Definition 7.1 involves a minimization over all leakage functions $\phi : \mathbb{F}_2^n \to \mathbb{R}$, which appears as an unsolvable task if no information about the underlying algorithm and leakage model is known [2]. In order to solve this issue, we choose a basis decomposition for the unknown *univariate* leakage function ϕ. Note that, $\phi : \mathbb{F}_2^n \to \mathbb{R}$ is a pseudo-Boolean function that can be seen as a 2^n-dimensional vector $\phi(\cdot) \in \mathbb{R}^{2^n}$. Here $\mathbb{R}^{2^n}$ is the usual finite-dimensional Euclidean space equipped with the canonical scalar product $\langle \phi \mid \phi' \rangle = \sum_x \phi(x)\phi'(x)$. The key point is that for the stochastic collision attack we can make use of the *full (canonical) basis*, i.e., basis of highest degree, as the distinguisher involves (at least) two leakage samples.

Lemma 7.1 *The stochastic collision attack of highest degree can be rewritten as follows:*

$$\mathscr{D}_{sto.coll}(\vec{x}^{(\cdot)}, \vec{l}^{(\cdot)}) = \underset{\vec{k} \in (\mathbb{F}_2^n)^W}{\operatorname{argmin}} \min_{\vec{a} \in \mathbb{R}^{2^n}} \sum_{q=1}^{N} \left\| \vec{l}_q - \sum_{u \in \mathbb{F}_2^n} a_u \delta_u(\vec{x}_q \oplus \vec{k}) \right\|^2, \quad (7.6)$$

where $\delta_u(x) = 1$ if $x = u$ and 0 otherwise.

Proof In Definition 7.1 the leakage model space can be mathematically described using an orthonormal basis in $\mathbb{R}^{2^n}$ (see, e.g., [9]). Using the canonical basis $(\delta_u)_{u \in \mathbb{F}_2^n}$, which is evidently orthonormal, simply gives $\phi(x) = \sum_{u \in \mathbb{F}_2^n} \phi(u)\delta_u(x)$. In Eq. (7.6), we denoted $\vec{a} = (\phi(u))_{u \in \mathbb{F}_2^n}$. □

Still, the stochastic collision distinguisher as presented in Eq. (7.6) is not satisfactory. Indeed, it is not a mathematical closed form, hence some optimization algorithm shall be run in order to minimize over $\vec{a} \in \mathbb{R}^{2^n}$. As the function inside the minimization (over $\vec{a}$) in Eq. (7.6) is actually convex, there is a global minimum, and a projection could hence be done as a first step of the optimization. However, the derivation of the global minimum implies a matrix inversion (see for instance [10,

---

[2] Note that we employ the notation $\operatorname{argmin}_{k^{(\cdot)}} \min_\phi$, which suggests that, first of all, a minimization on ϕ is performed for a set of vectorial keys $k^{(\cdot)}$, and then that the minimal value for all $k^{(\cdot)}$ is sought, and the corresponding key set is returned. But actually, the minimization is not forced to be in this order, as $k^{(\cdot)} \in (\mathbb{F}_2^n)^W$ and $\phi : \mathbb{F}_2^n \to \mathbb{R}$ are independent variables.

Sect. 4.2]), which is computationally intensive, and suffers drawbacks about accuracy and situations where the rank is not full. As a consequence of these observations, it would be preferable to derive the distinguisher by computing mathematically the value of the minimum (instead of directly running optimization algorithms). In the following subsection, we prove that this derivation can be done and we exhibit the closed form.

7.2.3 Main Result

This theorem is the main result of [5]:

Theorem 7.1 *The stochastic collision attack defined in Eq. (7.6) is equal to*

$$\mathcal{D}_{sto.coll}(\vec{x}^{(\cdot)}, \vec{l}^{(\cdot)}) = \underset{\vec{k} \in (\mathbb{F}_2^n)^W}{argmix} \sum_{u \in \mathbb{F}_2^n} \frac{\left(\sum_w \sum_{q/x_q^{(w)} \oplus k^{(w)} = u} l_q^{(w)}\right)^2}{\sum_w \sum_{q/x_q^{(w)} \oplus k^{(w)} = u} 1}. \tag{7.7}$$

In Eq. (7.7), "$\sum_{q/x_q^{(w)} \oplus k^{(w)} = u}$" is short for "$\sum_{\substack{q \in \{1,\dots,N\} \\ s.x.\ x_q^{(w)} \oplus k^{(w)} = u}}$".

Proof Let us denote by $G_{k^{(w)}}^{(w)}$ the $N \times 2^n$ matrix, where $G_{k^{(w)}}^{(w)}[q, u] = \delta_u(x_q^{(w)} \oplus k^{(w)})$. We use $\| \cdot \|$ to define the norm-2 in $\mathbb{R}^N$. Then, Eq. (7.6) can be expressed as

$$\mathcal{D}_{sto.coll}(\vec{x}^{(\cdot)}, \vec{l}^{(\cdot)}) = \underset{\vec{k} \in (\mathbb{F}_2^n)^W}{argmin} \ \underset{\vec{a} \in \mathbb{R}^{2^n}}{min} \ \sum_{w=1}^{W} \left\| \vec{l}^{(w)} - G_{k^{(w)}}^{(w)} \vec{a} \right\|^2. \tag{7.8}$$

In this equation, $\vec{a}$ is seen as a column of 2^n real numbers and $G_{k^{(w)}}^{(w)} \vec{a}$ is a matrix vector product. Now, we can solve the minimization on $\vec{a} \in \mathbb{R}^{2^n}$ by minimizing the convex function:

$$\sum_w \left\| \vec{l}^{(w)} - G_{k^{(w)}}^{(w)} \vec{a} \right\|^2 = \sum_w \left\| \vec{l}^{(w)} \right\|^2 \tag{7.9}$$

$$+ \vec{a}^\mathsf{T} \left(\sum_w G_{k^{(w)}}^{(w)\mathsf{T}} G_{k^{(w)}}^{(w)} \right) \vec{a} - 2\vec{a}^\mathsf{T} \left(\sum_w G_{k^{(w)}}^{(w)\mathsf{T}} \vec{l}^{(w)} \right), \tag{7.10}$$

where $\vec{a}^\mathsf{T}$ is the vector (i.e., the line-matrix of dimension 1×2^n) equal to the transposed of $\vec{a}$. This quadratic form is minimal when its Jacobian is equal to zero [11]. Let us denote:

- $\Sigma_{k^{(\cdot)}}$ the square $2^n \times 2^n$ matrix $\sum_w G_{k^{(w)}}^{(w)\mathsf{T}} G_{k^{(w)}}^{(w)}$ and
- $b_{k^{(\cdot)}}$ the column of 2^n elements $\sum_w G_{k^{(w)}}^{(w)\mathsf{T}} \vec{l}^{(w)}$.

Let us assume $\Sigma_{k(\cdot)}$ is invertible in the following, then we have $\vec{a} = \left(\Sigma_{k(\cdot)}\right)^{-1} b_{k(\cdot)}$. The optimal value of Eq. (7.10) is

$$\sum_w \left\| \vec{l}^{(w)} \right\|^2 + b_{\vec{k}}^{\mathsf{T}} \Sigma_{\vec{k}}^{-1} \Sigma_{\vec{k}} \Sigma_{\vec{k}}^{-1} b_{\vec{k}} - 2 b_{\vec{k}}^{\mathsf{T}} \Sigma_{\vec{k}}^{-1} b_{\vec{k}}$$

$$= \mathrm{cst} - b_{\vec{k}}^{\mathsf{T}} \left(\Sigma_{\vec{k}}\right)^{-1} b_{\vec{k}} \ , \tag{7.11}$$

because $\Sigma_{\vec{k}}$ is symmetrical (hence its inverse is also). Let apart the constant, the term $-b_{\vec{k}}^{\mathsf{T}} \left(\Sigma_{\vec{k}}\right)^{-1} b_{\vec{k}}$ is negative, because $\left(\Sigma_{\vec{k}}\right)^{-1}$ is positive definite. Thus, in the sequel, we rather focus on maximizing the opposite of this value.

Let us now develop the expressions. For each coordinate of $b_{\vec{k}} = (b_{\vec{k}}[u])_{u \in \mathbb{F}_2^n}$ we have:

$$b_{\vec{k}}[u] = \sum_{q=1}^{N} \sum_w G_{k^{(w)}}^{(w)}[q, u] l_q^{(w)}$$

$$= \sum_{q=1}^{N} \sum_w \delta_u (x_q^{(w)} \oplus k^{(w)}) l_q^{(w)}$$

$$= \sum_w \sum_{q / x_q^{(w)} \oplus k^{(w)} = u} l_q^{(w)} \ .$$

Besides, the element $(u, v) \in \mathbb{F}_2^n \times \mathbb{F}_2^n$ of the square matrix $\Sigma_{\vec{k}}$ can be rewritten as

$$\Sigma_{\vec{k}}[u, v] = \sum_{q=1}^{N} \sum_w \delta_u (x_q^{(w)} \oplus k^{(w)}) \, \delta_v (x_q^{(w)} \oplus k^{(w)})$$

$$= \begin{cases} \sum_w \sum_{q / x_q^{(w)} \oplus k^{(w)} = u} 1 & \text{if } u = v, \\ 0 & \text{otherwise.} \end{cases}$$

Thus, $\Sigma_{\vec{k}}$ is diagonal as well as its inverse; every element of the diagonal shall simply be inversed. Therefore, the objective function (Eq. (7.11)) takes a closed form expression. Finally, the stochastic collision distinguisher rewrites

$$\mathscr{D}_{\text{sto.coll}}(\vec{x}^{(\cdot)}, \vec{l}^{(\cdot)}) = \operatorname*{argmix}_{\vec{k} \in (\mathbb{F}_2^n)^W} \sum_{u \in \mathbb{F}_2^n} \left(\Sigma_{\vec{k}}[u, u]\right)^{-1} b_{\vec{k}}[u]^2$$

$$= \operatorname*{argmix}_{\vec{k} \in (\mathbb{F}_2^n)^W} \sum_{u \in \mathbb{F}_2^n} \frac{\left(\sum_w \sum_{q / x_q^{(w)} \oplus k^{(w)} = u} l_q^{(w)}\right)^2}{\sum_w \sum_{q / x_q^{(w)} \oplus k^{(w)} = u} 1} \ .$$

$\square$

The stochastic collision distinguisher can be computed efficiently thanks preliminary accumulation of traces. This optimization is also known as coalescence. The detailed description of the computation of $\mathscr{D}_{\text{sto.coll}}(\vec{x}^{(\cdot)}, \vec{l}^{(\cdot)})$ is given in Algorithm 7.1.

Algorithm 7.1 Stochastic collision distinguisher computed efficiently with coalescence

---

input : $\vec{x}^{(\cdot)}$ and $\vec{l}^{(\cdot)}$, a series of W plaintexts and leakages
output: $\mathscr{D}_{\text{sto.coll}}(\vec{x}^{(\cdot)}, \vec{l}^{(\cdot)}) \in (\mathbb{F}_2^n)^W$, as defined in Eq. (7.7).

// Coalescence of input traces ... avg[.,.] $\leftarrow \{0\} \in \mathbb{R}^{2^n \times W}$ // A $2^n \times W$ matrix of reals

1 count[.,.] $\leftarrow \{0\} \in \mathbb{N}^{2^n \times W}$ // A $2^n \times W$ matrix of integers
2 **for** $q \in \{1, \ldots, N\}$ **do**
3   **for** $w \in \{1, \ldots, W\}$ **do**
4   avg$[x_q^{(w)}, w] \leftarrow$ avg$[x_q^{(w)}, w] + l_q^{(w)}$ count$[x_q^{(w)}, w] \leftarrow$ count$[x_q^{(w)}, w] + 1$

// Maximum likelihood estimation   this_$\vec{k} \leftarrow 0 \in (\mathbb{F}_2^n)^W$ this_$\mathscr{D} \leftarrow 0 \in \mathbb{R}^+$ **for**
$\vec{k}^{(\cdot)} \in (\mathbb{F}_2^n)^W$ **do**
5   avg[.] $\leftarrow \{0\} \in \mathbb{R}^{2^n}$ // A vector of 2^n reals
6   count[.] $\leftarrow \{0\} \in \mathbb{N}^{2^n}$ // A vector of 2^n integers
7   **for** $u \in \mathbb{F}_2^n$ **do**
8   **for** $w \in \{1, \ldots, W\}$ **do**
9   avg$[u] \leftarrow$ avg$[u]$ + avg$[u, w]$ count$[u] \leftarrow$ count$[u]$ + count$[u, w]$

10     $\mathscr{D} \leftarrow 0$ **for** $u \in \mathbb{F}_2^n$ **do**
11        $\mathscr{D} \leftarrow \mathscr{D} +$ avg$[u]^2/$count$[u]$
12   **if** $\mathscr{D} >$ this_$\mathscr{D}$ **then**   // Found a better candidate
13   this_$\vec{k} \leftarrow \vec{k}^{(\cdot)}$ this_$\mathscr{D} \leftarrow \mathscr{D}$

---

14 **return** this_$\vec{k}$ // $\mathscr{D}_{\text{sto.coll}}(\vec{x}^{(\cdot)}, \vec{l}^{(\cdot)}) = \text{argmax}_{\vec{k} \in (\mathbb{F}_2^n)^W}$ $\sum_{u \in \mathbb{F}_2^n} \dfrac{\left(\sum_w \sum_{q/x_q^{(w)} \oplus k^{(w)} = u} l_q^{(w)} \right)^2}{\sum_w \sum_{q/x_q^{(w)} \oplus k^{(w)} = u} 1}$

// Notice: here, we assume for instance that $k^{(0)}$ is known or
brute-forced

7.3 Conclusion

The coalescence principle leads to significant complexity reductions (time and memory) in several SCAs. These improvements happen, asymptotically, without any loss of effectiveness. Furthermore, it turns the physical leakage into a pseudo-Boolean function (a real-value function defined over the Z's values). Thereby several more mathematical tools can be used on, among them the Walsh-Hadamard Transform (WHT) and the Numerical Normal Form (NNF) [12]. These tools could lead to further improvements.

References

1. Ouladj M, El Mrabet N, Guilley S, Guillot P, Millérioux G (2020) On the power of template attacks in highly multivariate context. J Cryptograph Eng - JCEN (2020)
2. Cochran WG (1934) The distribution of quadratic forms in a normal system, with application to the analysis of covariance. Math Proc Cambridge Philos Soc 30:178–191
3. Heuser A, Rioul O, Guilley S, Good is not good enough - deriving optimal distinguishers from communication theory. In: Batina and Robshaw [14], pp 55–74
4. Prouff E, Rivain M, Bevan R (2009) Statistical analysis of second order differential power analysis. IEEE Trans Comput 58(6):799–811
5. Bruneau N, Carlet C, Guilley S, Heuser A, Prouff E, Rioul O (2017) Stochastic collision attack. IEEE Trans Inf Forens Secur 12(9):2090–2104
6. Bruneau N, Guilley S, Heuser A, Rioul O (2014) Masks will fall off – higher-order optimal distinguishers. In: Sarkar P, Iwata T (eds), Advances in cryptology – ASIACRYPT 2014 - 20th international conference on the theory and application of cryptology and information security, Kaoshiung, Taiwan, R.O.C., December 7-11, 2014, Proceedings, Part II. Lecture notes in computer science, vol 8874. Springer, pp 344–365
7. Bruneau N, Guilley S, Heuser A, Marion D, Rioul O, Less is more - dimensionality reduction from a theoretical perspective. In: Güneysu and Handschuh [99], pp 22–41
8. Mahalanobis PC (1936) On the generalised distance in statistics. Proc Natl Inst Sci India 2(1):49–55
9. Werner S, Kerstin L, Paar C (2005) A model stochastic, for differential side channel cryptanalysis. In: LNCS (ed), CHES. LNCS, vol 3659. Springer, Edinburgh, pp. 30–46
10. François-Xavier S, François K, Werner S (2009) How to compare profiled side-channel attacks? In: Applied cryptography and network security. Lecture notes in computer science, vol 5536. Springer, Berlin, pp 485–498
11. Beveridge GSG, Schechter RS (1970) Optimization: theory and practice. Includes indexes. McGraw-Hill, New York
12. Carlet C, Guillot P (1999) A new representation of Boolean functions. In: Fossorier MPC, Imai H, Lin S, Poli A (eds), AAECC. Lecture notes in computer science, vol 1719, pp 94–103. Springer

Linear Regression Analysis with Coalescence Principle

8

8.1 Introduction

Since the seminal paper on Side-Channel Attacks (SCA) by Kocher et al. [1], several improvements have been published. As showed in Sect. 3.4 the most efficient SCA to date is the Template Attack (TA) [2]. This method is split into two phases that is a profiling and a matching stage. An important weakness of this attack is the large number of measurements required during the profiling stage. This requirement is due to the need of an accurate estimation of the probability densities of different instances of leakage [3]. Another limitation of TA is that it requires the access to an open copy of the target device to conduct the profiling phase with known secret parameters. To overcome these inconveniences, Schindler introduced a new stochastic approach, that consists in approximating probability densities [4]. In fact, even if the stochastic approach is less efficient than the Template Attack (TA), it has the advantage that it requires less traces in the profiling stage than TA [3] and that it can be turned into non-profiled attacks [5]. Since the publication of [4], several researches studied the stochastic approach as an SCA [6,7].

8.1.1 State-of-the-Art's Review

In the original paper published by Schindler et al. [4], the LRA is presented with a profiling step, as an alternative to the Template Attack [2,8]. In [8], the authors state the ability of the stochastic attack to "learn" quickly from a smaller number of samples than the Template Attack. A profiling attack using multivariate regression analysis, followed by a Pearson correlation coefficient processing, is presented in [9]. The authors of this paper have shown empirically that this approach is more efficient than the conventional stochastic model attack.

In [5], it has been shown that the LRA can be carried out without a profiling stage. In this paper, authors argued that the LRA can be conducted in the univariate context like the Correlation Power Analysis (CPA) [10]. The advantage of this suggestion is

© The Author(s), under exclusive license to Springer Nature Switzerland AG 2021
M. Ouladj and S. Guilley, *Side-Channel Analysis of Embedded Systems*,
https://doi.org/10.1007/978-3-030-77222-2_8

that the LRA needs less assumptions than CPA. The relationship between the CPA and the stochastic approach is studied in [11].

Recently, LRA applied to XOR operations has been studied using the Hamming weight and the distance models by the authors of [12]. They found that in many common scenarios, LRA is a more efficient tool than CPA.

Subsequently, extensions of LRA to protected implementations are studied in [3, 13, 14]. Using the later paper, authors of [15] address the question of the LRA efficiency processing. This chapter provides a near exponential improvement of the LRA computing time and an exponential improvement of memory space compared with [15]. Our improvements apply to the LRA either with or without a profiling phase, knowing the subkey during the profiling phase or not.

Recently, the authors of [16] suggested to use the Walsh-Hadamard orthogonal basis to characterize the leakage. We improve the processing time on this particular basis. Our improvements are held independently from the basis.

8.1.2 Contributions

Let us assume that N observations $(\vec{l_i})_{0 \le i \le N-1}$ (D-dimensional traces) have been measured during the processing of an operation $F(x_i, k)$ where F and the x_i are assumed to be known and where k is a target secret assumed to be constant. Moreover, let $\mathscr{L}$ denote the $N \times D$-matrix whose rows are the $\vec{l_i}$'s. The stochastic approach essentially consists in modeling, for each key candidate k, the leakage $\mathscr{L}$ according to a basis parameterized by k. The core idea is that a good choice of the basis should imply that the modeling achieves best accuracy when the key guess k is correct.

Let us denote by $\mathscr{M}_k$ the $N \times s$-matrix whose rows correspond to the evaluation of the s basis functions in the values $F(x_i, k)$. All of the previously published papers suggest to compute the matrix $\mathscr{P}_k \doteq \left(\mathscr{M}_k{}^\mathsf{T} \mathscr{M}_k\right)^{-1} \mathscr{M}_k{}^\mathsf{T}$ in order to carry out the coefficients of the leakage on the chosen basis. Then $\mathscr{M}_k \mathscr{P}_k \mathscr{L}$ is computed as an estimator of the leakage (or directly $(\mathscr{M}_k \mathscr{P}_k) \mathscr{L}$) as shown in Algorithm 8.2. This processing is the bottleneck of the LRA processing, because it is repeated for all guessed subkeys k.

As above, let us denote by n the bit-length of k and x_i. Even though the authors of [14, 15] give some computation improvements, the complexity of their algorithm is $\mathscr{O}(2^{3n})$ in the space memory and $\mathscr{O}(2^{3n} \times D)$ in the processing time. Thanks to the duo *coalescence and spectral approach*, this chapter provides further significant improvements in both of them; the required memory space becomes $\mathscr{O}(s \times 2^n)$ and the processing time reduced to $\mathscr{O}(sn2^n d)$. In what follows, we summarize our contributions.

First, an invariant computation of matrices is exhibited (the matrix $\left(\mathscr{M}_k{}^\mathsf{T} \mathscr{M}_k\right)^{-1}$ is independent from the guessed subkey k), thereby avoiding $2^n - 1$ repeated computations. In particular, only one matrix inversion is required during the whole attack, which leads to a significant reduction in the processing time.

Second, a relationship between stochastic model expressions for the guessed subkey 0 and an arbitrary guessed subkey k is exhibited (namely, between $\mathcal{M}_0\mathcal{P}_0$ and $\mathcal{M}_k\mathcal{P}_k$). This relationship allows for drastic improvements both in the pre-processing time and in the memory space required to carry out an LRA. Thanks to this relation, only one product matrix is needed for all possible subkeys k.

Third, we exhibit a spectral approach to further accelerate LRA computations (Algorithm 8.3 instead of Algorithm 8.2). This allows for a near exponential computational improvement (a time attack factor is reduced from 2^n to n). For reference, the use of the spectral approach in SCAs was introduced in [17]. In what follows, we provide a further improvement of ratio $2^n/s$ both in the space memory and the processing time (Algorithm 8.4 instead of Algorithm 8.3).

Fourth, we extend those improvements in the higher order masking context (Sect. 8.4). For our experimental validation, two masking methods are challenged, namely, the Boolean masking and the 2^n-modular additive masking [18].

Finally, we exhibit, for the first time the relationship between the LRA and the Numerical Normal Form (NNF) [19] of the leakage function. In fact, in order to carry out a higher order (H-O) SCA, an optimal combination between the shares leakages must be found (see [20] for more details). If the noise variance is high, the optimal combination is the *normalized product*, independently from the masking technique [21]. Thanks to the relation between the NNF and the LRA, we show that it is possible to exhibit the polynomial expression of this combination against the first-order additive masking. In fact, knowing such a polynomial expression is useful for the H-O CPA. Also, knowing the degree of the polynomial is useful for the adversary to choose the basis used in the LRA.

8.1.3 Outline

In Sect. 8.2, a mathematical modelization of the problem is presented. In Sect. 8.3, we develop our contributions in a monovariate context (where the sensitive variable is not shared). In Sect. 8.4, we extend them to multivariate contexts (i.e., when masking countermeasure is applied). In Sect. 8.5, we verify our contributions on simulated AES traces. Finally, we conclude our study in Sect. 8.6.

8.2 Mathematical Modelization

8.2.1 Description of Stochastic Attacks

LRA consists first in choosing a basis of s functions $(m_p)_{0\leq p\leq s-1}$, such that m_0 is a constant function (typically equal to 1). For each subkey guess k, the adversary constructs an $(N \times s)$−*prediction matrix* $\left(\mathcal{M}_k[i][p] \doteq m_p(z_i)\right)_{i,p}$, such that $z_i = F(x_i, k)$ is the guessed value of the sensitive variable corresponding to the plaintext x_i. For example, if one aims at regressing the leakage as a linear combination of the bits of the z_i's, then for this basis one has $s = 9$, $m_0(z) = 1$ and $m_p(z)$ returning the pth bit of z.

The comparison of this matrix with the set of D-dimensional leakage traces $(\vec{l_i})_i \hookleftarrow \vec{L}$ is carried out by linear regression of each coordinate of $(\vec{l_i})$ in the basis formed by the rows of $\mathcal{M}_k$. Namely, a real-valued $(s \times D)$-matrix β_k with column vectors $\vec{\beta_k}[1], \ldots, \vec{\beta_k}[D]$ are estimated in order to minimize the error when approximating $\mathcal{L}[u]$ by $(\mathcal{M}_k \times \beta_k)[u]$. The vector $\vec{\beta_k}[u]$ is defined, for each time sample u and each guess k, such that [22, Theorem 3]:

$$\vec{\beta_k}[u] = (\mathcal{M}_k{}^\mathsf{T}\mathcal{M}_k)^{-1}\mathcal{M}_k{}^\mathsf{T}\vec{\mathcal{L}}[u], \tag{8.1}$$

where $\mathcal{L}$ is a $(N \times D)$−matrix, whose rows are the $\vec{l_i}{}^\mathsf{T}$ vectors, and where $\vec{\mathcal{L}}[u]$ denotes the uth column of the matrix $\mathcal{L}$.

As defined in [15], to quantify the estimation error, the *goodness of fit model* is used and the *correlation coefficient of determination* R is computed for each u. R is defined by $R \doteq 1 - SSR/\overrightarrow{SST}$, where $\overrightarrow{SST}$ denotes the leakage *Total Sum of Squares* and SSR denotes the leakage *Residual Sum of Squares* [15].

Let us remind that the dimension of $\mathcal{M}_k$ is $N \times s$. So, carrying out LRA naively on the real traces is difficult and even impossible in presence of countermeasures (a large number N of measurements is required) with high dimension D. One can show that the attack efficiency is unchanged if it is performed over the averaged leakages per class (the averaged leakage equals $(\frac{1}{\#i:x_i=x}\sum_{i:x_i=x}\vec{l_i})_{x\in\mathbb{F}_2^n}$, if $(\#i : x_i = x) \neq 0$, and it equals 0 else) instead of over rough traces (as suggested in [14] and argued in the precedent chapter). Thanks to this improvement, the size of $\mathcal{M}_k$ is reduced from $\mathcal{O}(N \times s)$ to $\mathcal{O}(2^n \times s)$. We recall that N (which is the number of traces involved in the attack, e.g., $N = 10^6$) is often much greater than 2^n (the number of possible subkeys, e.g., $2^n = 256$).

Algorithm 8.2 shows the LRA processing on the averaged leakages per sensitive variable value, as suggested in the literature ([14, 22]). As we can see from this algorithm, for each guessed subkey k, the attack can be conducted first by preprocessing $\mathcal{P}_k = (\mathcal{M}_k{}^\mathsf{T}\mathcal{M}_k)^{-1}\mathcal{M}_k{}^\mathsf{T}$ once, then by computing the weights of linear regression $\vec{\beta_k}[u] = \mathcal{P}_k \times \vec{\mathcal{L}}[u]$ for each u, and eventually by computing the leakage estimation $\vec{\mathcal{E}_k}[u] = \mathcal{M}_k\mathcal{P}_k \times \vec{\mathcal{L}}[u]$. Remember that the overall complexity of the inversion is about $\mathcal{O}(s^{2.373})$ using Optimized Coppersmith–Winograd-like algorithm [23]. So, the time complexity of the pre-processing of the 2^n matrices $\mathcal{P}_k$ is about $\mathcal{O}(2^n \times (s^{2.373} + 2^n \times s^2))$ and that of the vectors $\vec{\beta_k}[u]$ is about $\mathcal{O}(2^{2n} \times s \times D)$ multiplications.

Since the adversary does not need the coefficients of $\vec{\beta_k}[u]$ explicitly, during a non-profiled attack, one can pre-process $\mathcal{G}_k = \mathcal{M}_k\mathcal{P}_k$ once, then straightforwardly compute the leakage estimation $\vec{\mathcal{E}_k}[u] = \mathcal{G}_k \times \vec{\mathcal{L}}[u]$, for each u. So, we can save one matrix product per (k, u) [15]. Whether using this optimization or not, the LRA is still difficult (even impossible against higher order masking). In fact, one must pre-compute and save 2^n matrices $\mathcal{G}_k$ and compute one matrices' product $\vec{\mathcal{E}_k}[u] = \mathcal{G}_k \times \vec{\mathcal{L}}[u]$ for each pair (k, u). So, the pre-processing time complexity of the 2^n

Algorithm 8.1 Linear Regression Analysis with coalescence (without using the Propositions 8.1 and 8.2)

---

input :
- a set of D-dimensional leakage $(\vec{l}_i)_{1 \leq i \leq N}$ and the corresponding plaintexts $(x_i)_{1 \leq i \leq N}$,
- a set of model functions $(m_p)_{p \leq s}$.

output: The candidate subkey k

1   **for** $i \leftarrow 0$ **to** $N - 1$ **do**   // Processing of the leakage Total Sum of Squares ($\overrightarrow{SST}$)

2      $\mu_{\vec{\mathscr{L}}} \leftarrow \mu_{\vec{\mathscr{L}}} + \vec{l}_i$

3      $v_{\vec{\mathscr{L}}} \leftarrow v_{\vec{\mathscr{L}}} + \vec{l}_i^2$

4   $\overrightarrow{SST} \leftarrow v_{\vec{\mathscr{L}}} - \frac{1}{N}\mu_{\vec{\mathscr{L}}}^2$

5   **for** $k \leftarrow 0$ **to** $2^n - 1$ **do**   // Pre-processing of the 2^n prediction matrices $\mathscr{M}_k$ and $\mathscr{P}_k$

6   **for** $p \leftarrow 0$ **to** s **do**

7   **for** $x \leftarrow 0$ **to** $2^n - 1$ **do**

8              $\mathscr{M}_k[x][p] \leftarrow m_p[F(x,k)]$

9      $\mathscr{P}_k \leftarrow \left(\mathscr{M}_k^{\mathsf{T}} \times \mathscr{M}_k\right)^{-1} \times \mathscr{M}_k^{\mathsf{T}}$

10   **for** $i \leftarrow 0$ **to** $N - 1$ **do**   // Computation of the coalesced matrix $\mathscr{L}$

11      $\mathscr{L}^{\mathsf{T}}[x_i] = \mathscr{L}^{\mathsf{T}}[x_i] + l_i$

12   count$[x_i]$=count$[x_i]$+1 **for** $p \leftarrow 0$ **to** s **do**

13   **for** $x \leftarrow 0$ **to** $2^n - 1$ **do**

14              $\mathscr{M}_k[x][p] \leftarrow m_p[F(x,k)]$

15      $\mathscr{P}_k \leftarrow \left(\mathscr{M}_k^{\mathsf{T}} \times \mathscr{M}_k\right)^{-1} \times \mathscr{M}_k^{\mathsf{T}}$

16   **for** $x \leftarrow 0$ **to** $2^n - 1$ **do**

17   **if** *count[x]* **then**

18          $\mathscr{L}^{\mathsf{T}}[x] = \mathscr{L}^{\mathsf{T}}[x]/\text{count}[x]$

19   **for** $u \leftarrow 0$ **to** $D - 1$ **do**   // Instantaneous attack (at time u)

20      $\vec{\beta} \leftarrow \mathscr{P}_k \times \vec{\mathscr{L}}[u]$

21      $\vec{\mathscr{E}}[u] \leftarrow \mathscr{M}_k \times \vec{\beta}$ // $\vec{\mathscr{E}}[u] = \mathscr{M}_k \times \left(\mathscr{M}_k^{\mathsf{T}} \times \mathscr{M}_k\right)^{-1} \times \mathscr{M}_k^{\mathsf{T}} \times \vec{\mathscr{L}}[u]$: Estimator of $\vec{\mathscr{L}}[u]$

22      $SSR \leftarrow 0$

23   **for** $x \leftarrow 0$ **to** $2^n - 1$ **do**   // Computation of the Residual Sum of Squares (SSR)

24          $SSR \leftarrow SSR + (\mathscr{E}[x][u] - \mathscr{L}[x][u])^2$

25      $R[k][u] \leftarrow 1 - \frac{SSR}{\overrightarrow{SST}[u]}$ // coefficient of determination [14]

26 **return** $\text{argmax}_k \left(max_u R[k][u]\right)$

matrices $\mathscr{G}_k$ is about $\mathscr{O}(s^{2.373} \times 2^n + 2^{2n} \times s^2 + 2^{3n} \times s)$ and that of the estimators $\hat{\mathscr{E}}_k[u]$ is about $\mathscr{O}(2^{3n} \times D)$ multiplications.

To circumvent these bottlenecks, an LRA study with some important improvements in terms of memory space and processing time will be provided in the rest of this chapter. These improvements concern both scenarios with and without processing of the regression weights $\vec{\beta}_k[u]$.

8.3   LRA Study and Improvements of Its Implementation

As previously recalled, in practice the LRA is carried out over the leakages averaged according to the value of the corresponding sensitive variable (or equivalently according to the value of the corresponding message). This approach was first put forward by [14,15,22] and is called "*coalesced*" in [24]. One of the advantages of averaging the leakage is that, as long as $N \geq 2^n$, it leads to less memory space and more processing time efficiency.

In fact, as studied in Chap.7, carrying out an SCA with or without averaging traces is asymptotically equivalent.

So, in the rest of this chapter we study the LRA with coalescence, hence assuming that the adversary does not directly attack using the N traces $\vec{l}_i$ but using their averaged (*coalesced*) version (recall that $n_j \doteq \#\{x_i = j; \ 0 \leq i \leq N - 1\}$, and the averaged trace equals $\frac{1}{n_j} \sum_{x_i=j}^{N} \vec{l}_i$, if $n_j \neq 0$, and it equals 0 else). To simplify the notations, the 2^n averaged traces are still denoted by $\vec{l}_i$ and the dimension N in the definition of the matrices $\mathscr{M}_k$, $\mathscr{P}_k$, $\mathscr{G}_k$ and $\mathscr{L}$ is replaced by 2^n. Namely, the matrix $\mathscr{M}_k$ dimension becomes $2^n \times s$ instead of $N \times s$ and the matrix $\mathscr{L}$ dimension becomes $2^n \times D$ instead of $N \times D$ and similarly for all the derived matrices and vectors. Those new matrices shall be called *coalesced* in the sequel.

8.3.1   LRA with Assumption of Equal Images Under different Subkeys (EIS)

An important property, that holds true for the most key mixing-based cryptographic functions, is introduced in [22]. Namely, it is the EIS property introduced by Definition 1 in the Chap. 6.

Recall that a more specific case of this property is when $F(x, k)$ depends only on $x \oplus k$ as in common block ciphers (i.e., on ciphers as DES [25] and AES [26]). Other cases are the 2^n-modular addition ($\mod (2^n)$), and the $2^{2^m} + 1$-modular multiplication of non-zero numbers, such that m=0,1,2,3,4 (in fact $(2^{2^m} + 1)_{m=0,1,2,3,4}$ are the first Fermat numbers).

The first important contribution of the current chapter is the following proposition.

Proposition 8.1 *Using the coalescence, let* $\mathscr{H} \doteq \mathscr{M}_k^{\mathsf{T}} \mathscr{M}_k$, $\mathscr{P}_k \doteq (\mathscr{M}_k^{\mathsf{T}} \mathscr{M}_k)^{-1} \mathscr{M}_k^{\mathsf{T}}$ *and* $\mathscr{G}_k \doteq \mathscr{M}_k \mathscr{P}_k$, *such that* $\mathscr{M}_k$ *is as defined above. Under the EIS assumption:*

1. $\mathscr{H}$ *is independent from the guessed subkey* k.
2. *If the leakage function depends only on* $x \oplus k$, *then for every* (j, k) *we have*

$$\mathscr{M}_k[j][.] = \mathscr{M}_0[j \oplus k][.]$$

and,

(a) $\mathscr{P}_k[i][j] = \mathscr{P}_0[i][j \oplus k]$.
(b) $\mathscr{G}_k[i][j] = \mathscr{G}_0[i \oplus k][j \oplus k]$.

Proof 1.   From the EIS assumption,

$$\mathscr{H}_k[i][j] = \sum_{x=0}^{2^n-1} \mathscr{M}_k^{\mathsf{T}}[i][x] \mathscr{M}_k[x][j] = \sum_{x=0}^{2^n-1} m_i[F(x,k)] m_j[F(x,k)]$$

$$= \sum_{y=0}^{2^n-1} m_i[F(y,0)] m_j[F(y,0)] = \mathscr{H}_0[i][j].$$

If the leakage only depends on $x \oplus k$, then

$$\mathscr{H}_k[i][j] = \sum_{x=0}^{2^n-1} \mathscr{M}_k^{\mathsf{T}}[i][x] \mathscr{M}_k[x][j] = \sum_{x=0}^{2^n-1} \mathscr{M}_0^{\mathsf{T}}[i][x \oplus k] \mathscr{M}_0[x \oplus k][j]$$

$$= \sum_{y=0}^{2^n-1} \mathscr{M}_0^{\mathsf{T}}[i][y] \mathscr{M}_0[y][j] = \mathscr{H}_0[i][j] \quad ; \text{such that } y = x \oplus k.$$

2. (a)

$$\mathscr{P}_k[i][j] = (\mathscr{H}^{-1} \times \mathscr{M}_k^{\mathsf{T}})[i][j] = \sum_{p=0}^{s-1} \mathscr{H}^{-1}[i][p] \mathscr{M}_k^{\mathsf{T}}[p][j]$$

$$= \sum_{p=0}^{s-1} \mathscr{H}^{-1}[i][p] \mathscr{M}_0^{\mathsf{T}}[p][j \oplus k] = \mathscr{P}_0[i][j \oplus k].$$

(b)

$$\mathscr{G}_k[i][j] = (\mathscr{M}_k \times \mathscr{P}_k)[i][j] = \sum_{p=0}^{s-1} \mathscr{M}_k[i][p] \mathscr{P}_k[p][j]$$

$$= \sum_{p=0}^{s-1} \mathscr{M}_0[i \oplus k][p] \mathscr{P}_0[p][j \oplus k] = \mathscr{G}_0[i \oplus k][j \oplus k].$$

◇   □

Proposition 8.1 implies a significant time and memory gain. First, only one matrix $\mathscr{H}_0$ has to be computed and inverted instead of 2^n.

Second, for each time sample u, if we need the weight $\vec{\beta}_k[u]$ returned by the linear regression, then we just need to first pre-process the single matrix $\mathscr{P}_0$. All the other $2^n - 1$ matrices $\mathscr{P}_k$ can be directly deduced using Proposition 8.1 instead of pre-processing and saving 2^n matrices products. Afterward, for every guessed subkey k and for each time sample u, one matrices' product needs to be carried out to get the coefficients $\vec{\beta}_k[u] = \mathscr{P}_k \vec{\mathscr{L}}[u]$.

Third, similarly to above, to conduct LRA without computing $\vec{\beta}_k[u]$ vectors, the adversary needs first to pre-process a single matrix $\mathscr{G}_0$. All the other $2^n - 1$ matrices $\mathscr{G}_k$ can be directly deduced using Proposition 8.1 instead of 2^n matrices products pre-processing and savings, as suggested in [15]. Then for every guessed subkey k and for each time sample u, one matrix product has to be carried out to compute $\vec{\mathscr{E}}_k[u] = \mathscr{G}_k \vec{\mathscr{L}}[u]$. In fact, whether this improvement is used or not, the estimator's computing is still difficult.

8.3.2 Spectral Approach Computation to Speed up LRA (with EIS)

In this section, we will show that the LRA processing can still be improved. Indeed, instead of computing coefficients $\vec{\beta}_k[u]$ (resp. vectors $\vec{\mathscr{E}}_k[u]$) for each subkey k, only one computation is sufficient, as shown in the sequel. Namely, thanks to convolution property, one can evaluate the convolution product in all possible values k in once, with an overall complexity of $\mathcal{O}(n2^n)$ instead of $\mathcal{O}(2^{2n})$ as explained in the Eq. 5.3.

Proposition 8.2 *With the same previous notations, if $F(x, k) = F(x \oplus k)$ for every possible values of x and k, then we have*

1. $\beta_k[i][u] = WHT^{-1}\left(WHT(\mathscr{P}_0[i][.]) \bullet WHT(\vec{\mathscr{L}}[u])\right)(k)$

2. $\mathscr{E}_k[i][u] = WHT^{-1}\left(WHT(\mathscr{G}_0[i \oplus k][.]) \bullet WHT(\vec{\mathscr{L}}[u])\right)(k)$,

where

- $\mathscr{P}_0[i][.]$ *denotes the ith line of the matrix $\mathscr{P}_0$,*
- $\vec{\mathscr{L}}[u]$ *denotes the uth column of the matrix $\mathscr{L}$.*

Proof 1.

$$\beta_k[i][u] = (\mathscr{P}_k \mathscr{L}[u])[i] = \sum_{j=0}^{2^n} \mathscr{P}_k[i][j] \mathscr{L}[j][u] = \sum_{j=0}^{2^n} \mathscr{P}_0[i][j \oplus k] \mathscr{L}[j][u]$$

$$= \mathscr{P}_0[i][.] \otimes \vec{\mathscr{L}}[u](k) = WHT^{-1}\left(WHT(\mathscr{P}_0[i][.]) \bullet WHT(\vec{\mathscr{L}}[u])\right)(k),$$

by seeing the line $\mathscr{P}_0[i][.]$ and the vector $\vec{\mathscr{L}}[u]$) as function evaluation tables.

2. Similarly,

$$\mathcal{E}_k[i][u] = (\mathcal{G}_k \vec{\mathcal{L}}[u])[i] = \sum_{j=0}^{2^n-1} \mathcal{G}_0[i \oplus k][j \oplus k]\mathcal{L}[j][u] = \mathcal{G}_0[i \oplus k][.] \otimes \vec{\mathcal{L}}[u](k)$$

$$= WHT^{-1}\left(WHT(\mathcal{G}_0[i \oplus k][.]) \bullet WHT(\vec{\mathcal{L}}[u])\right)(k).$$

◇   □

So, Proposition 8.2 allows further save time. Indeed if we need the $\vec{\beta}_k[u]$ coefficients, we first need to pre-process only one matrix $\mathcal{P}_0$. Then we compute $WHT(\mathcal{P}_0[i][.])$ for each line $0 \leq i \leq s - 1$, only once. Then we also compute $WHT(\vec{\mathcal{L}}[u])$ only once. Then for each line $0 \leq i \leq s - 1$, we compute the vector $WHT^{-1}\left(WHT(\mathcal{P}_0[i][.]) \bullet WHT(\vec{\mathcal{L}}[u])\right)$. In fact, according to Proposition 8.2, for each k the coefficient $\vec{\beta}_k[i][u]$ is the kth scalar of the last vector. So, no matrix product should be done to compute the coefficient $\beta_k[i][u]$. The overall time complexity of the pre-processing is about $\mathcal{O}(s^{2.373} + 2^n \times s^2 + s \times n2^n)$ and that of vectors $\vec{\beta}_k[u]$ computation is about $\mathcal{O}((n + 1)2^n \times s \times D)$. So, the vectors' $\vec{\beta}_k[u]$ computation has an overall complexity of $\mathcal{O}((n + 1)2^n \times s \times D)$ instead of $\mathcal{O}(2^{2n} \times s \times D)$.

Similarly, to conduct LRA without computing the $\vec{\beta}_k[u]$ vectors, the adversary needs first to pre-process only one matrix $\mathcal{G}_0$. Then she pre-processes a WHT for each line $0 \leq i \leq 2^n - 1$ of $\mathcal{G}_0$. Then she modifies the leakage vector $\vec{\mathcal{L}}[u]$ by $WHT(\vec{\mathcal{L}}[u])$. Then she modifies each line $0 \leq i \leq 2^n - 1$ of $\mathcal{G}_0$ by the vector $WHT^{-1}(WHT(\mathcal{G}_0[i][.]) \bullet WHT(\vec{\mathcal{L}}[u]))$ (only a coordinate-wise product and a processing of WHT^{-1} to do per line). Finally, for each guessed subkey k and each plaintext $0 \leq i \leq 2^n - 1$, the estimated leakage $\mathcal{E}_k[i][u]$ is just a scalar of the updated matrix $\mathcal{G}_0$. Namely it is the current value of $\mathcal{G}_0[i \oplus k][k]$. So, no matrix product must be done during the estimated leakage matrix $\mathcal{E}_k$ computation. The overall time complexity of the pre-processing is about $\mathcal{O}(s^{2.373} + 2^{n+1} \times s^2 + n2^{2n})$ and that of the estimated leakage matrix $\mathcal{E}_k$ computation is about $\mathcal{O}((n2^n + (n + 1)2^{2n})D)$. So, the estimated leakage matrix's $\mathcal{E}_k$ computation has an overall complexity of $\mathcal{O}((n + 1)2^{2n} \times D)$ instead of $\mathcal{O}(2^{3n} \times D)$. Algorithm 8.3 shows LRA processing with this spectral approach.

8.3.3   Further Improvement

The new bottleneck of the last algorithm (Algorithm 8.3) is the lines 19–20. To avoid computing 2^n Walsh-Hadamard Transform inverse WHT^{-1}, one can see that lines from 19 to 24 aim to compute the error square quadratic norm $\left\|\vec{\mathcal{L}}[u] - \vec{\mathcal{E}}_k[u]\right\|^2$. In

---

Algorithm 8.2 Linear Regression Analysis with coalescence and the spectral approach

---

input :
- a set of D-dimensional leakage $(\vec{l_i})_{0 \leq i \leq N-1}$ and the corresponding plaintexts $(x_i)_{0 \leq i \leq N-1}$,
- a set of model functions $(m_p)_{p \leq s}$. •

output: The candidate subkey k

27 **for** $i \leftarrow 0$ **to** $N-1$ **do**   // Processing of the leakage Total Sum of
      Squares $(\overrightarrow{SST})$

28 $\quad \mu_{\vec{\mathscr{L}}} \leftarrow \mu_{\vec{\mathscr{L}}} + \vec{l_i}$

29 $\quad v_{\vec{\mathscr{L}}} \leftarrow v_{\vec{\mathscr{L}}} + \vec{l_i}^2$

30 $\overrightarrow{SST} \leftarrow v_{\vec{\mathscr{L}}} - \frac{1}{N}\mu_{\vec{\mathscr{L}}}^2$

31 **for** $p \leftarrow 0$ **to** s **do**   // Pre-processing of the unique prediction matrix
      $\mathscr{M}_0$

32 $\quad$ **for** $x \leftarrow 0$ **to** $2^n - 1$ **do**

33 $\quad\quad \mathscr{M}_0[x][p] \leftarrow m_p[F(x,0)]$

34 $\mathscr{G}_0 \leftarrow \mathscr{M}_0 \times \left(\mathscr{M}_0^{\mathsf{T}} \times \mathscr{M}_0\right)^{-1} \times \mathscr{M}_0^{\mathsf{T}}$

35 **for** $i \leftarrow 0$ **to** $2^n - 1$ **do**   // Pre-processing WHT of the lines of matrix $\mathscr{G}_0$

36 $\quad \mathscr{G}_0[i][.] \leftarrow WHT(\mathscr{G}_0[i][.])$

37 **for** $i \leftarrow 0$ **to** $N-1$ **do**   // Computation of the coalesced matrix $\vec{\mathscr{L}}$

38 $\quad \vec{\mathscr{L}}^{\mathsf{T}}[x_i] = \vec{\mathscr{L}}^{\mathsf{T}}[x_i] + l_i$

39 $\quad$ count$[x_i]$=count$[x_i]$+1

40 **for** $x \leftarrow 0$ **to** $2^n - 1$ **do**

41 $\quad$ **if** $count[x]$ **then**

42 $\quad\quad \vec{\mathscr{L}}^{\mathsf{T}}[x] = \vec{\mathscr{L}}^{\mathsf{T}}[x]/\text{count}[x]$

43 **for** $u \leftarrow 0$ **to** $D-1$ **do**   // Instantaneous attack (at time u)

44 $\quad \vec{W}_L[u] \leftarrow WHT(\vec{\mathscr{L}}[u])$

45 $\quad$ **for** $x \leftarrow 0$ **to** $2^n - 1$ **do** // Computation of the estimator $\vec{\mathscr{E}}[u]$ according
      to Proposition 8.2

46 $\quad\quad \vec{\mathscr{E}}[x][.] \leftarrow WHT^{-1}\left(\mathscr{G}_0[x][.] \bullet \vec{W}_L[u]\right)$

47 $\quad$ **for** $k \leftarrow 0$ **to** $2^n - 1$ **do**   // Test hyp. k for all leakage coordinates

48 $\quad\quad SSR \leftarrow 0$

49 $\quad\quad$ **for** $x \leftarrow 0$ **to** $2^n - 1$ **do**   // Computation of the Residual Sum of
        Squares (SSR)

50 $\quad\quad\quad SSR \leftarrow SSR + (\mathscr{E}[x \oplus k][k] - \mathscr{L}[x][u])^2$

51 $\quad\quad R[k][u] \leftarrow 1 - \frac{SSR}{\overrightarrow{SST}[u]}$ // coefficient of determination [14]

52 **return** argmax$_k$ $(max_u R[k][u])$

---

fact, the LRA aims to minimize this error. One can minimize it otherwise:

$$\left\|\vec{\mathscr{L}}[u] - \vec{\mathscr{E}}_k[u]\right\|^2 = \sum_{i=0}^{2^n-1} (\mathscr{L}[i][u] - \mathscr{E}_k[i][u])^2$$

$$= \sum_{i=0}^{2^n-1} \mathscr{L}[i][u]^2 + \sum_{i=0}^{2^n-1} \mathscr{E}_k[i][u]^2 - 2\sum_{i=0}^{2^n-1} \mathscr{L}[i][u]\mathscr{E}_k[i][u].$$

First, $\sum_{i=0}^{2^n-1} \mathscr{L}[i][u]^2$ is clearly independent from k.

Second, since the symmetric matrix $\mathscr{G}_k$ is idempotent (i.e., $\mathscr{G}_k{}^\mathsf{T}\mathscr{G}_k = \mathscr{G}_k$), so

$$\sum_{i=0}^{2^n-1} \mathscr{E}_k[i][u]^2 = \sum_{i=0}^{2^n-1} (\mathscr{G}_k\vec{\mathscr{L}}[u])[i]^2 = \sum_{i=0}^{2^n-1} ((\mathscr{G}_k\vec{\mathscr{L}}[u])^\mathsf{T}[i]) \times ((\mathscr{G}_k\vec{\mathscr{L}}[u])[i])$$

$$= (\mathscr{G}_k\vec{\mathscr{L}}[u])^\mathsf{T} \times ((\mathscr{G}_k\vec{\mathscr{L}}[u])) = \vec{\mathscr{L}}[u]^\mathsf{T}\mathscr{G}_k{}^\mathsf{T}\mathscr{G}_k\vec{\mathscr{L}}[u] = \vec{\mathscr{L}}[u]^\mathsf{T}\mathscr{G}_k\vec{\mathscr{L}}[u]$$

$$= \sum_{i=0}^{2^n-1} \mathscr{L}[i][u]\mathscr{E}_k[i][u].$$

Finally, we have $\left\|\vec{\mathscr{L}}[u] - \vec{\mathscr{E}}_k[u]\right\|^2 = \sum_{i=0}^{2^n-1}\mathscr{L}[i][u]^2 - \sum_{i=0}^{2^n-1}\mathscr{L}[i][u]\mathscr{E}_k[i][u].$

So, minimizing the estimation error is nothing but maximizing $\sum_{i=0}^{2^n-1}\mathscr{L}[i][u]\mathscr{E}_k[i][u]$.

Recall that $\mathscr{G}_k = \mathscr{M}_k(\mathscr{M}_k{}^\mathsf{T}\mathscr{M}_k)^{-1}\mathscr{M}_k{}^\mathsf{T}$. Since $(\mathscr{M}_k{}^\mathsf{T}\mathscr{M}_k)$ is a positive defined matrix, one can factorize it by Cholesky's decomposition into a matrix product $\mathscr{T}^\mathsf{T}\mathscr{T}$, such that $\mathscr{T}$ is an $s \times s$ upper triangular matrix [27]. So, let us define by $\mathscr{U}_k$ the matrix $\mathscr{T}^{\mathsf{T}-1}\mathscr{M}_k{}^\mathsf{T}$. Then, we have $\mathscr{G}_k = \mathscr{U}_k{}^\mathsf{T}\mathscr{U}_k$ and,

$$\sum_{i=0}^{2^n-1} \mathscr{L}[i][u]\mathscr{E}_k[i][u] = \vec{\mathscr{L}}[u]^\mathsf{T}\mathscr{G}_k\vec{\mathscr{L}}[u] = (\mathscr{U}_k\vec{\mathscr{L}}[u])^\mathsf{T}(\mathscr{U}_k\vec{\mathscr{L}}) = \left\|\mathscr{U}_k\vec{\mathscr{L}}[u]\right\|^2.$$

So, minimizing the error square quadratic norm it's equivalent to maximizing $\left\|\mathscr{U}_k\vec{\mathscr{L}}[u]\right\|^2$. Moreover, one can show that $\mathscr{U}_k[.][j] = \mathscr{U}_0[.][j \oplus k]$. Hence, we have

$$\left\|\mathscr{U}_k\vec{\mathscr{L}}[u]\right\|^2 = \sum_{p=0}^{s-1} \left((\mathscr{U}_k\vec{\mathscr{L}}[u])[p]\right)^2 = \sum_{p=0}^{s-1} \left(\sum_{j=0}^{2^n-1} \mathscr{U}_0[p][j \oplus k]\mathscr{L}[j][u]\right)^2$$

$$= \sum_{p=0}^{s-1} \left(\mathscr{U}_0[p][.] \otimes \vec{\mathscr{L}}[u](k)\right)^2$$

$$= \sum_{p=0}^{s-1} \left(WHT^{-1}\left(WHT\left(\mathscr{U}_0[p][.]\right) \bullet WHT(\vec{\mathscr{L}}[u])\right)(k)\right)^2$$

So, $\left\|\mathscr{U}_k\vec{\mathscr{L}}[u]\right\|^2 = \left\|WHT^{-1}\left(WHT\left(\mathscr{U}_0[.][.]\right) \bullet WHT(\vec{\mathscr{L}}[u])\right)(k)\right\|^2,$ (8.2) such that $WHT(\mathcal{U}_0[.][.])$ denotes WHT for each line of the matrix $\mathcal{U}_0$.

Finally, one can carry out the LRA by saving only one matrix $\mathcal{U}_0$ of dimension $s \times 2^n$, then computing $WHT(\mathcal{U}_0[.][.])$, and processing for each u the convolution of each line with the leakage vector according to Eq. (8.2). The obtained key by the LRA is the column index that has the highest norm in the result matrix. One can easily sow that the *correlation coefficient of determination* $R \doteq 1 - SSR/\overrightarrow{SST}$ is nothing but $\left\| \mathcal{U}_k \vec{\mathcal{L}}[u] \right\|^2 / \overrightarrow{SST}$.

Algorithm 8.4 shows the most efficient LRA processing that we have obtained, thanks to (1) the coalescence, (2) the last transformation using $\mathcal{U}_k$, and (3) the spectral approach. In fact Algorithm 8.4 is $2^n/s$ times more efficient than Algorithm 8.3 both in space memory and processing time. Tables 8.1 and 8.2 summarize the LRA best complexities, with and without computing $\vec{\beta}_k[u]$ vectors, respectively. The third lines give our improvement in the AES case of study ($n = 8$) for basis size $s = 9$.

It is noticeable that all our results that assume the group operation $\oplus$ over the set $\{0, 1\}^n$ (*i.e.*, $F(x, k) = F(x \oplus k)$) hold true for any other group operation $\odot$ as long as Walsh-Hadamard Transform is replaced by Fourier Transform (FFT) on this group ($\{0, 1\}^n, \odot$). For example, if the operation is $+(\mod(2^n))$, then Cyclical Fourier Transform must replace Walsh-Hadamard Transform (WHT).

It is also noteworthy that to increase the numerical stability of the least square estimation, authors of [3, 16] advise to select an orthogonal basis. Another advantage of our improvements is that is independent from the chosen basis.

Furthermore, our improvements relate to the LRA either with or without a profiling phase, either knowing the subkey during the profiling phase or not.

Table 8.1 Space memory and executing time complexities with computing $\vec{\beta}$

	Memory	$\mathscr{P}$ Pre-processing	$\vec{\beta}$ Processing
Old implementations	$s \times 2^{2n}$	$s^{2.373} \times 2^n + 2^{2n} \times s^2$	$2^{2n} \times s \times D$
Our implementation	$s \times 2^n$	$s^{2.373} + 2^n \times s^2 + s \times n2^n$	$(n + 1)2^n \times s \times D$
AES ($n = 8$) with $s = 9$	256 times	136 times	28 times

Table 8.2 Space memory and executing time complexities without computing $\vec{\beta}$

	Memory	Pre-processing	$\vec{\mathscr{E}}$ Processing
Old implementations	2^{3n}	$s^{2.373} \times 2^n + 2^{2n} \times s^2 + 2^{3n} \times s$	$2^{3n} \times D$
Our implementation	$s \times 2^n$	$s^{2.373} + 2^{n-1} \times s^2 + s \times n2^n$	$(n + s(n + 1))2^n D$
AES ($n = 8$) with $s = 9$	7282 times	5426 times	736 times

Algorithm 8.3 The most efficient LRA thanks to the spectral approach

input :
- a set of D-dimensional leakage $(\vec{l}_i)_{0 \le i \le N-1}$ and the corresponding plaintexts $(x_i)_{0 \le i \le N-1}$,
- a set of model functions $(m_p)_{p \le s}$.

output: The candidate subkey k

53 **for** $i \leftarrow 0$ **to** $N-1$ **do**   // Processing of the leakage Total Sum of
   Squares $(\overrightarrow{SST})$
54 | $\mu_{\vec{\mathscr{L}}} \leftarrow \mu_{\vec{\mathscr{L}}} + \vec{l}_i$
55 | $v_{\vec{\mathscr{L}}} \leftarrow v_{\vec{\mathscr{L}}} + \vec{l}_i^{\,2}$
56 $\overrightarrow{SST} \leftarrow v_{\vec{\mathscr{L}}} - \frac{1}{N}\mu_{\vec{\mathscr{L}}}^2$
57 **for** $p \leftarrow 0$ **to** s **do**   // Pre-processing of the unique prediction matrix
   $\mathscr{M}_0$
58 | **for** $x \leftarrow 0$ **to** $2^n - 1$ **do**
59 | | $\mathscr{M}_0[x][p] \leftarrow m_p[F(x,0)]$

60 $\mathscr{U}_0 \leftarrow \mathscr{T}^{\mathsf{T}^{-1}}\mathscr{M}_0^{\mathsf{T}}$ // $\mathscr{T} \doteq \left(Cholesky\left(\mathscr{M}_0^{\mathsf{T}} \times \mathscr{M}_0\right)\right)^{-1}$
61 **for** $p \leftarrow 0$ **to** s **do**   // Pre-processing WHT of the lines of matrix $\mathscr{U}_0$
62 | $\mathscr{U}_0[p][.] \leftarrow WHT(\mathscr{U}_0[p][.])$

63 **for** $i \leftarrow 0$ **to** $N-1$ **do**   // Computation of the coalesced matrix $\vec{\mathscr{L}}$
64 | $\vec{\mathscr{L}}^{\mathsf{T}}[x_i] = \vec{\mathscr{L}}^{\mathsf{T}}[x_i] + l_i$
65 | $count[x_i] = count[x_i] + 1$
66 **for** $x \leftarrow 0$ **to** $2^n - 1$ **do**
67 | **if** $count[x]$ **then**
68 | | $\vec{\mathscr{L}}^{\mathsf{T}}[x] = \vec{\mathscr{L}}^{\mathsf{T}}[x]/count[x]$

69 **for** $u \leftarrow 0$ **to** $D-1$ **do**   // Instantaneous attack (at time u)
70 | $\vec{W}_L[u] \leftarrow WHT(\vec{\mathscr{L}}[u])$
71 | **for** $p \leftarrow 0$ **to** $s-1$ **do**   // Computation of the error according to
   | Eq. (8.2)
72 | | $\mathscr{U}_0[p][.] \leftarrow WHT^{-1}\left(\mathscr{U}_0[p][.] \bullet \vec{W}_L[u]\right)$
73 | **for** $k \leftarrow 0$ **to** $2^n - 1$ **do**   // Test hyp. k for all leakage coordinates
74 | | $SSR \leftarrow 0$
75 | | **for** $p \leftarrow 0$ **to** $s-1$ **do**   // Computation of the Residual Sum of
   | | Squares (SSR)
76 | | | $SSR \leftarrow SSR + (\mathscr{U}_0[p][k])^2$
77 | | $R[k][u] \leftarrow \frac{SSR}{\overrightarrow{SST}[u]}$ // coefficient of determination [14]

78 **return** $\text{argmax}_k\left(max_u R[k][u]\right)$

8.3.4 Incremental Implementation of LRA

Incremental implementation of LRA means the carrying out of an LRA using some set of traces, then update its computation when a new trace is added, and so on. From Algorithm 8.3 one can see that the instructions 5–10 should be carried out only once. All the other instructions can easily be implemented incrementally according to the new added leakage traces. It is useful to note that the instructions 18–20 can also be implemented incrementally, due to the linearity of WHT and the bi-linearity of the coordinate-wise product ($\bullet$).

8.4 Extension of the Improvements to the Protected Implementations by Masking

Let us consider a leaking cryptographic device protected by *masking*. Let d denotes a strictly positive integer. The protection by a masking of order d consists in dividing each sensitive variable Z into $d + 1$ shares $Z^{(0)},..., Z^{(d)}$. In practice at least two different kinds of masking can be used; the Boolean masking and the arithmetic masking [18]. While the Boolean masking is compatible with linear Boolean functions, the arithmetic masking is compatible with arithmetic operations such as the modular addition in IDEA and TEA Algorithms [28]. Without loss of generality, we assume that the u-th sample in the measured leakage $\vec{L}$ corresponds to the u-th share of the sensitive variable Z, such that $0 \leq u \leq d$. Let the deterministic part of the leakage $\vec{L}$ (namely $\mathbb{E}(\vec{L}|(Z^{(u)})_{0 \leq u \leq d})$) be denoted by $\vec{M}$. As shown in Sect. 3.3.1, In order to recover information on Z, an adversary must combine the leakages on all the $d + 1$ shares in a so-called $(d + 1)$-th higher order SCA.

In order to conduct a higher order attack, many combination functions (which aim at combining the leakages of the different shares to form a new exploitable signal) were suggested in the literature [29–31]. According to [21,32], for a high noise, the optimal combination technique against masking is the *normalized* product (for a first-order masking: $(\vec{\mathscr{L}}[0] - \mathbb{E}(\vec{L}[0]))(\vec{\mathscr{L}}[1] - \mathbb{E}(\vec{L}[1])))$, and this combination function should be accompanied by an optimal model (for a first-order masking: $\mathbb{E}((\vec{M}_k[0] - \mathbb{E}(\vec{M}_k[0]))(\vec{M}_k[1] - \mathbb{E}(\vec{M}_k[1]))|\ Z))$. While the LRA is carried out on the combination leakage, as suggested in [14], all the ideas developed in the previous sections to improve the LRA efficiency can straightforwardly be applied on the protected implementations. In Sect. 8.5.3, we experimentally show through several examples that the effectiveness of LRA against masked implementations stays unchanged when applying our improvements (the timing and the memory complexities are significantly improved).

8.4.1 Normalized Product Combination Against Arithmetic Masking

Since the construction of a proper SCA leakage model is at least as important as the selection of a good distinguisher [5], in this subsection we assess the optimal leakage model of the multiplication combination against arithmetic masking.

Recall that in practice the n-variables polynomials' basis is the most used basis in the LRA-based SCAs. In such a case, LRA consists in projecting the leakage function $\vec{\mathscr{L}}^{\mathsf{T}}[.]$ on the polynomials' basis. Our core idea is that for any function f defined from $\{0, \ldots 2^n - 1\}$ to $\mathbb{R}$ such that for each $Z \in \{0, \ldots 2^n - 1\}$ the value $f(Z)$ is known, the Numerical Normal Form (NNF) [19] consists to project f also on this polynomials' basis. So, in H-O attacks knowing the polynomial projection of the model (or at least knowing its degree), the (normalized) multiplication combination allows the adversary to choose the optimal polynomials' basis for LRA. In fact, if one uses the polynomials' basis, which is the most used in LRA, the expected coefficients $\vec{\beta}_k[u]$ are none other than the coefficients of the Numerical Normal Form (NNF) [19] of $\vec{\mathscr{E}}[u]$. That means LRA does nothing but regresses the NNF of the Leakage function when a polynomials' basis is used.

Thanks to [33, Lemma 1], one knows that if the leakage follows the Hamming weight model, then the combination against the dth-order Boolean masking is a degree 1 polynomial. Namely it equals $(-\frac{1}{2})^d \left(\sum_i z_i - \frac{n}{2}\right)$, such that $(z_i)_{i=0\ldots7}$ denotes the bits of the sensitive value Z. This important result holds for the Boolean masking but it does not hold for the arithmetic masking.

For the first-order modular addition masking, let the combination function $f \doteq \mathbb{E}((HW(Z + R) - \mathbb{E}(HW(Z + R)))(HW(R) - \mathbb{E}(HW(R))) \mod 256|\ Z)$. In fact thanks to the Numerical Normal Form of the function f denoted $\widetilde{f}$, one can show that it is a degree 2 polynomial of Z's bits. Namely $\widetilde{f} = 4608 + \sum_i (2^i - 256)z_i + \sum_{i,j; i<j} (2^{8+i-j} z_i z_j)$. Similarly, if the leakage follows the Hamming weight model, the optimal combination of d shares ($d < 9$) for such masking is a dth-degree polynomial. To exhibit this result, one should compute $f(Z)$ for each possible value Z, then carry out the NNF of f using Algorithm 8.4 obtained from [19].

Algorithm 8.4 Numerical Normal Form [19]

input : f, n.
output: f as its $\widetilde{f}$

```
79  for i ← 0 to n − 1 do
80      b ← 0
81      repeat
82          for x ← b to b + 2^i − 1 do
83              ⌊ f[x + 2^i] ← f[x + 2^i] − f[x]
84          b ← b + 2^{i+1}
85      until b = 2^n
```

8.5 Experiments

During our experiments two simulated leakage models are studied by

1. The Hamming weight model: $\sum_{z_i=0}^{7} z_i$,
2. A degree 2 polynomial model: $\sum_{z_i=0}^{7} z_i + \frac{1}{2} \sum_{z_i=0}^{6} \sum_{z_j=z_i+1}^{7} z_i z_j$.

In both of them we added a zero-mean additive Gaussian noise, with different standard deviation values σ. In this work we compare CPA with coalescence, CPA, LRA with coalescence and LRA as long as its carrying out is possible. All these attacks are targeting the $s - box$ output of the first AES round.

We assess different attacks both against unprotected implementations and against implementations protected by the Boolean masking or arithmetic masking. In the different scenarios, we assume that, with the exception of the additive independent noise, the CPA's adversary knows the leakage model. The effectiveness of our attacks is assessed by processing Success Rates (SR) according to the numbers of traces (#measurements) [34].

8.5.1 LRA with and Without Spectral Approach

First, we assess the efficiency of LRA with and without the spectral approach. Figure 8.1 shows that there is no difference between them in terms of success rate, in all scenarios (protected/unprotected implementations with 1 or 2 degree polynomial leakage model). So, thanks to our improvements, one can conduct LRA without loss in the effectiveness, while it is significantly faster, with less memory space (as shown theoretically in Tables 8.1, 8.2 and experimentally in Tables 8.3, 8.4).

Table 8.3 Space memory and executing time of LRA with computing $\vec{\beta}$; AES use case

	Memory	$\mathscr{P}$ pre-processing	$\vec{\beta}$ processing
Old implementations	589824 floats	58×10^{-3} s	8.297×10^{-3} s
Our implementation	2304 floats	$0,445 \times 10^{-3}$ s	0.332×10^{-3} s
AES ($n = 8$) with $s = 9$	256 times	130 times	25 times

Table 8.4 Space memory and executing time of LRA without computing $\vec{\beta}$; AES use case

	Memory	Pre-processing	$\vec{\mathscr{E}}$ processing
Old implementations	16777216 floats	1.63 s	0.236 s
Our implementation	2304 floats	0.329×10^{-3} s	0.333×10^{-3} s
AES ($n = 8$) with $s = 9$	7282 times	4954 times	709 times

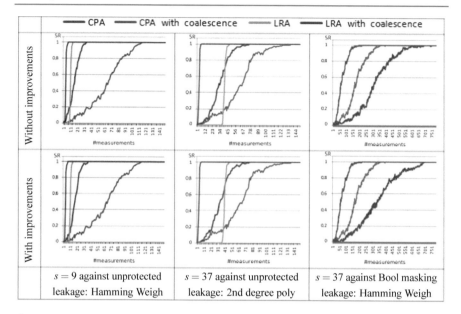

Fig. 8.1 Success rate of CPA and LRA (with/without our implementation improvements) according to the number of measurements, such that $\sigma = 0$

8.5.2 SCAs with and Without Coalescence

The coalescence approach leads to more efficient attacks in terms of computing time. Figure 8.1 shows that there is a big difference between attacks with and without coalescence, in terms of success rate. In fact, as shown in Sect. 7.1 and explained in Sect. 7.1, the coalescence approach requires to have asymptotically the same number n_i of messages x_i per class. This assumption does not hold if the attack requires few leakages/messages, that is the case for these figures ($\sigma = 0$).

In order to show the asymptotic equivalence between the SCAs with and without coalescence, one can study their success rates for high covariance noise. Figures 8.2 and 8.3 show this asymptotic equivalence when the noise is high.

8.5.3 LRA Against Higher Order Masking

To study the efficiency of the SCA attacks against masking (especially LRA with our computing improvements), comparisons between the first-order Boolean and the arithmetic masking (more precisely the addition modulo 2^8) are carried out on the AES S-box. Even if the modular addition masking is not useful for AES S-box, we study its efficiency over it in order to compare both masking techniques against same attacks on the same S-box. In these attacks we use the normalized multiplicative combination.

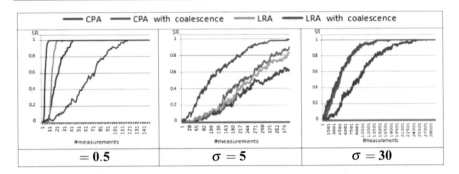

Fig. 8.2 Success rate of CPA and LRA (with/without coalescence) with basis size 9, against unprotected implementation, according to the number of measurements, such that the leakage is simulated to the Hamming weight model

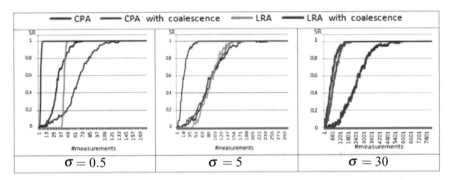

Fig. 8.3 Success rate of CPA and LRA (with our implementation improvements) with basis size 37, against unprotected implementation, according to the number of measurements, such that the leakage is simulated to degree 2 polynomial model

As shown in Figs. 8.4 and 8.5, and as explained in Sect. 8.4.1, the 9-size basis is suitable for the multiplication combination against Boolean masking. One can see that, against Boolean masking, the LRA is more efficient if the 9-size basis is used (right of Fig. 8.4) than that if the 37-size basis is used (right of Fig. 8.5).

In contrast, for the modular addition arithmetic masking, the suitable basis is a 37-size basis, which corresponds to degree 2 polynomial leakage function. One can see that, against arithmetic masking, LRA is less efficient if the 9-size basis is used (left of Fig. 8.4) than that if the 37-size basis is used (left of Fig. 8.5). Furthermore, Fig. 8.5 shows that the optimal leakage model of the normalized multiplicative combination is $4608 + \sum_i (2^i - 256)z_i + \sum_{i,j;i<j} (2^{8+i-j}z_iz_j)$ as exhibited in Sect. 8.4.1, thanks to the NNF. It labeled by *optimal CPA* in Fig. 8.5.

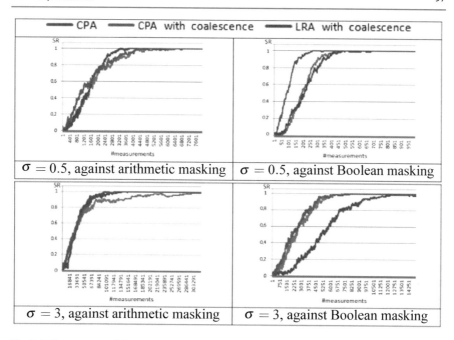

Fig. 8.4 Success rate of CPA and LRA (with our implementation improvements) with basis size 9, against first-order masking implementation, according to the number of measurements, such that the leakage is simulated to the Hamming weight model

8.6 Conclusion and Perspectives

In this chapter we studied the use of coalescence both in LRA, in CPA, and in covariance distinguishers. We also show that coalescence can accelerate the computation of LRA distinguisher, even in presence of masking countermeasure. Our study provides breakthrough improvements both in memory space and in running time required to carry out the LRA. Furthermore we exhibit the polynomial form of the normalized product combination which is useful in SCA against masking, especially for LRA. It is the first time when LRA is related to the Numerical Normal Form (NNF).

We believe that this work democratizes the use of LRA and that it leads to further research results in the field of parametric side-channel attacks.

In practice the leaking information can take place in many time samples. From an adversarial point of view, it is therefore beneficial to attempt to make the most of highly multivariate traces. Thereby, in the following chapter, it will be shown that the winning duo "coalescence principle and spectral approach" can also be applied in the multidimensional analysis (Template attacks as a use case), even against the masking-based protected implementations.

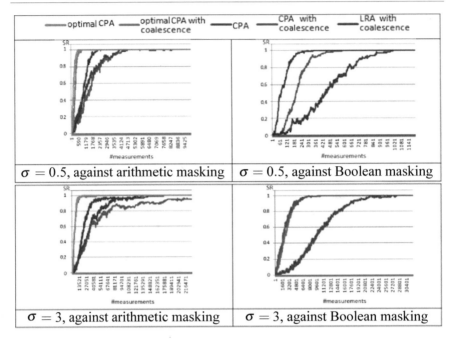

Fig. 8.5 Success rate of CPA and LRA (with our implementation improvements) with basis size 37, against 1st order masking implementation, according to the number of measurements, such that the leakage is simulated to the Hamming weight model

References

1. Kocher PC, Jaffe J, Jun B (1999) Differential power analysis. In: Proceedings of the 19th annual international cryptology conference on advances in cryptology, CRYPTO '99. Springer, London, pp 388–397

2. Chari S, Rao JR, Rohatgi P (2002) Template attacks. In: Kaliski BS Jr, Koç ÇK, Paar C (ed) Cryptographic hardware and embedded systems - CHES 2002, 4th international workshop, redwood shores, CA, USA, August 13-15, 2002, Revised papers. Lecture notes in computer science, vol 2523. Springer, pp 13–28

3. Werner S, (2008) Advanced stochastic methods in side channel analysis on block ciphers in the presence of masking. J Math Cryptol 2(3), 291–310, (2008) ISSN (Online) 1862–2984. ISSN (Print) 1862–2976: https://doi.org/10.1515/JMC.2008.013

4. Schindler W (2005) On the optimization of side-channel attacks by advanced stochastic methods. In: Vaudenay S (ed), Public key cryptography - PKC 2005, 8th international workshop on theory and practice in public key cryptography, Les Diablerets, Switzerland, January 23–26, 2005, Proceedings. Lecture notes in computer science, vol 3386. Springer, pp 85–103

5. Doget J, Prouff E, Rivain M, Standaert F-X (2011) Univariate side channel attacks and leakage modeling. J. Cryptograph Eng 1(2):123–144

6. Schaub A, Schneider E, Hollender A, Calasans V, Jolie L, Touillon R, Heuser A, Guilley S, Rioul O (2014) Attacking suggest boxes in web applications over HTTPS using side-channel

stochastic algorithms. In: Lopez J, Ray I, Crispo B (eds), Risks and security of internet and systems - 9th international conference, CRiSIS 2014, Trento, Italy, August 27–29, 2014, Revised selected papers. Lecture notes in computer science, vol 8924. Springer, pp 116–130

7. Bruneau N, Carlet C, Guilley S, Heuser A, Prouff E, Rioul O (2017) Stochastic collision attack. IEEE Trans Inf Forens Secur 12(9):2090–2104

8. Gierlichs B, Lemke-Rust K, Paar C (2006) Templates vs. stochastic methods. In: CHES. LNCS, vol 4249. Springer, Yokohama, pp 15–29

9. Sugawara T, Homma N, Aoki T, Satoh A (2010) Profiling attack using multivariate regression analysis. IEICE Electron Express 7(15):1139–1144

10. Brier É, Clavier C, Olivier F (2004) Correlation power analysis with a leakage model. In: Joye M, Quisquater J-J (eds), Cryptographic hardware and embedded systems - CHES 2004: 6th international workshop Cambridge, MA, USA, August 11–13, 2004. Proceedings. Lecture notes in computer science, vol 3156. Springer, pp 16–29

11. De Santis F, Kasper M, Mangard S, Sigl G, Stein O, Stöttinger M (2013) On the relationship between correlation power analysis and the stochastic approach: an ASIC designer perspective. In: Paul G, Vaudenay S (ed), Progress in cryptology - INDOCRYPT 2013 - 14th international conference on cryptology in India, Mumbai, India, December 7–10, 2013. Proceedings. Lecture notes in computer science, vol 8250. Springer, pp 215–226

12. Shan F, Wang Z, Wei F, Guoai X, Wang A (2017) Linear regression side channel attack applied on constant xor. IACR Cryptology ePrint Archive 2017:1217

13. Lemke-Rust K, Paar C (2007) Analyzing side channel leakage of masked implementations with stochastic methods. In: Computer security - ESORICS 2007, 12th European symposium on research in computer security, Dresden, Germany, September 24-26, 2007, Proceedings, pp 454–468

14. Dabosville G, Doget J, Prouff E (2013) A new second-order side channel attack based on linear regression. IEEE Trans Comput 62(8):1629–1640

15. Lomné V, Prouff E, Roche T (2013) Behind the scene of side channel attacks. In: Sako K, Sarkar P (eds), ASIACRYPT (1). Lecture notes in computer science, vol 8269. Springer, pp 506–525

16. Guilley S, Heuser A, Tang M, Rioul O (2017) Stochastic side-channel leakage analysis via orthonormal decomposition. In: Farshim P, Simion E (eds), Innovative security solutions for information technology and communications - 10th international conference, SecITC 2017, Bucharest, Romania, June 8–9, 2017, Revised selected papers. Lecture notes in computer science, vol 10543. Springer, pp 12–27

17. Guillot P, Millérioux G, Dravie B, El Mrabet N, Spectral approach for correlation power analysis. In: Hajji et al. [100], pp 238–253

18. Kerstin L, Kai S, Paar C (2004) DPA, on n-bit sized Boolean and arithmetic operations and its application to IDEA, RC6, and the HMAC-construction. In: CHES. Lecture notes in computer science, vol 3156. Springer, Cambridge, pp 205–219

19. Carlet C, Guillot P (1999) A new representation of Boolean functions. In: Fossorier MPC, Imai H, Lin S, Poli A (eds), AAECC. Lecture notes in computer science, vol 1719. Springer, pp 94–103

20. Lomné V, Prouff E, Rivain M, Roche T, Thillard A, How to estimate the success rate of higher-order side-channel attacks. In: Batina and Robshaw [14], pp 35–54

21. Bruneau N, Guilley S, Heuser A, Rioul O (2014) Masks will fall off – higher-order optimal distinguishers. In: Sarkar P, Iwata T (eds), Advances in cryptology – ASIACRYPT 2014 - 20th international conference on the theory and application of cryptology and information security, Kaoshiung, Taiwan, R.O.C., December 7–11, 2014, Proceedings, Part II. Lecture notes in computer science, vol 8874. Springer, pp 344–365

22. Werner S, Kerstin L, Paar C, A model stochastic, for differential side channel cryptanalysis. In LNCS, (ed), CHES. LNCS, vol 3659. Springer, Edinburgh, pp 30–46

23. Williams VV (2012) Multiplying matrices faster than coppersmith-winograd. In: STOC'12 Proceedings of the forty-fourth annual ACM symposium on theory of computing, New York, USA — May 19 - 22, 2012, pp 887–898

24. Ouladj M, El Mrabet N, Guilley S, Guillot P, Millérioux G (2020) On the power of template attacks in highly multivariate context. J Cryptograph Eng - JCEN (2020)

25. NIST/ITL/CSD. Data Encryption Standard. FIPS PUB 46-3, Oct 1999. http://csrc.nist.gov/publications/fips/fips46-3/fips46-3.pdf

26. NIST/ITL/CSD. Advanced Encryption Standard (AES). FIPS PUB 197, Nov 2001. http://nvlpubs.nist.gov/nistpubs/FIPS/NIST.FIPS.197.pdf (also ISO/IEC 18033-3:2010)

27. Krishnamoorthy A, Menon D (2013) Matrix inversion using cholesky decomposition. In: 2013 signal processing: algorithms, architectures, arrangements, and applications (SPA) (2013), pp 70–72. ISBN:978-83-62065-17-2, INSPEC accession number: 14041759, Electronic ISSN: 2326-0319, Print ISSN: 2326-0262

28. Wheeler DJ, Needham RM (1994) Tea, a tiny encryption algorithm. In: Preneel B (ed) Fast software encryption: second international workshop. Leuven, Belgium, 14–16 December 1994, Proceedings. Lecture notes in computer science, vol 1008. Springer, pp 363–366

29. Schramm K, Paar C (2006) Higher order masking of the AES. In: Pointcheval D (ed), CT-RSA. LNCS, vol 3860. Springer, pp 208–225

30. Joye M, Paillier P, Schoenmakers B (2005) On second-order differential power analysis. In: CHES. LNCS, vol 3659. Springer, August 29 – September 1st 2005. Edinburgh, UK, pp 293–308

31. Oswald E, Mangard S (2007) Template attacks on masking — resistance is futile. In: Abe M (ed), CT-RSA. Lecture notes in computer science, vol 4377. Springer, pp 243–256

32. Prouff E, Rivain M, Bevan R (2009) Statistical analysis of second order differential power analysis. IEEE Trans Comput 58(6):799–811

33. Rivain M, Prouff E, Doget J (2009) Higher-order masking and shuffling for software implementations of block ciphers. In: CHES. Lecture notes in computer science, vol 5747. Springer, Lausanne, pp 171–188

34. Standaert F-X, Malkin T, Yung M (2009) A unified framework for the analysis of side-channel key recovery attacks. In: EUROCRYPT, vol 5479 of LNCS, pp 443–461. Springer, April 26-30 2009. Cologne, Germany

Template Attack with Coalescence Principle

9

On the one hand, template attacks have been introduced to deal with multivariate leakages, with as few assumptions as possible on the leakage model. On the other hand, many works have underlined the need for dimensionality reduction. In this chapter, we clarify the relationship between template attacks in full space and in linear subspaces.

In particular and thanks to "coalescence principle & spectral approach", we exhibit a clear mathematical expression for template attacks, which enables an efficient computation even on large dimensions such as several hundred samples.

9.1 Introduction

9.1.1 Context: The Side-Channel Threat

Side-channel traces collected from software code are extremely rich, since the same variable can leak at different places. Typically, leakage can spread over several samples within one clock cycle, and in addition, software implementations typically move variables in several registers or memory locations, causing leakage at many clock cycles.

9.1.2 Problem: Making the Most of High Dimensionality

Modern oscilloscopes sample their input at a very high frequency, hence, it is possible to get more than one leakage sample per leaking sensitive variable. How to exploit such an abundance of leakage measurements? Few non-supervised side-channel

© The Author(s), under exclusive license to Springer Nature Switzerland AG 2021
M. Ouladj and S. Guilley, *Side-Channel Analysis of Embedded Systems*,
https://doi.org/10.1007/978-3-030-77222-2_9

distinguishers manage such situation. Indeed, the samples usually leak differently, therefore, it is complex without prior knowledge to know how to best combine them constructively.

9.1.3 State-of-the-Art

Clearly, in the past, some strategies have been put forward. For instance, Clavier et al. [31, Sect. 3] suggested (albeit in another context, namely that of so-called shuffling countermeasures) to take as monovariate signal the average of the signal over the D samples. As another example, Hajra and Mukhopadhyay [1] investigate non-profiling side-channel analysis. However, for their analysis to work, some a priori structure is injected in the model, which is learned online on the attacking traces. However, in reality, the information contents of the whole trace are larger than that soaked from the projection on one single support vector, but it depends on the multivariate trace distribution. Notice that template attacks (which are profiled attacks) achieve this goal. Template attacks on multivariate data have been described accurately as a "process" in state-of-the-art papers [2, Sect. 5.3.3]. However, those attacks are still perceived as a recipe (see Sect. 2.1 of [3]), so there is no (except experimental cases) way to study which parameters impact the success of a template attack. Therefore, some folklore surrounds them. In particular, because the recipe is not formalized, some papers have tried to clarify the different steps and assumptions, in particular, Choudary and Kuhn [4] wrote on making many details about template attacks more explicit, such as the notion of "pooled covariance matrix". The inversion of the traces covariance matrix Σ is feared, and for this reason, the first pass of dimensionality reduction has been suggested right away in the seminal paper by [3]; the authors selected a few tens of so-called "Points of Interest" (PoIs), resulting in a smaller Σ which is easy to inverse. Still, we notice a theoretical contradiction, because the *data processing inequality* states that application of any function on the data reduces their informative contents. Subsequently, many researches have been carried out to improve on the heuristic method to identify PoIs in the traces. Archambeau et al. motivated the usefulness of Principal Component Analysis (PCA) in [5] in this respect. Bär et al. improved on this PCA in [6] by actually hand-picking PoIs within PCA vectors. Elaabid et al. [7] observed that choosing PoIs as all samples of PCA larger than a threshold actually retain most informative samples without the need of manual selection. Fan et al. suggest to select PoIs as those where the noise is the most Gaussian [8]. Still, even deciding which selection of PoIs is optimal is questioned [9]. Zhang et al. noticed recently that there is still a margin for improvement in PoIs selection algorithms [10]. So, to summarize, many papers have focused on reducing the traces dimensionality while retaining in the constructed subset most of the information, with a view to optimize the success rate. A secondary objective is to keep an acceptable computational load for the template attack. Indeed, template attacks, as presented originally, include in the attack phase the evaluation of computationally challenging functions, such as exponential, matrix inversion, square roots, etc. (see Eq. (1) in Sect. 2.1 of [3]).

Eventually, we notice that many papers study trace denoising, using typically wavelets [11], Independent Component Analysis (ICA) [12], etc. In our work, we exploit traces "raw", so as to highlight the sole impact of multivariate analysis on attack efficiency. But, it is noticeable that both PoI-based and PCA-based template attacks [8] can also straightforwardly benefit from our approach, both in term of processing time and in term of the needed space memory, even against the masking-based protected implementations.

9.1.4 Contributions

In this chapter, we analyze template attacks from a theoretical perspective and derive the mathematical expressions for template building and attack phases. In the context of multivariate normal noise, the two phases are easily simplified (Algorithms 9.6 and 9.7) as mere linear operations. In particular, there is no need for exponential and square roots, and only one matrix inversion is required at the end of the template building phase. These formal expressions allow us to drastically improve on the clarity of the actual computations carried out by template attacks and as a result, the formal expression also improves on the computational aspects related to template attacks.

Our noteworthy contributions are detailed below:

- we factor code between training and matching algorithms,
- we optimize further the computation and needed memory space by grouping traces by classes, resulting in computation on a "coalesced" data,
- we have the training phase delivers only one matrix, thereby saving repeated computations in the attack phase (see Algorithms 9.8 and 9.9).
- As a consequence, we manage to perform template attacks on highly multivariate traces. Contrary to belief, we show that the more samples are taken (i.e., no dimensionality reduction is needed), the more successful is the template attack. But our approach is also compatible with reduced dimensionality data.
- Said differently, we are able to analyze full-width traces without any pre-processing (which destroys information) and show that this is the best attack strategy, in terms of the number of traces, to recover the key.
- in the sequel, we extend our approach to the masking-based protected implementations.
- Finally, we highlight a spectral approach which allows for a near exponential computational improvement in the attack phase (an attack factor in 2^n is reduced to n, where n is the bitwidth of the sensitive variable).

Furthermore, those findings allow us to observe practically the effect of the dimensionality (the number of samples in a trace) on the success rate. Namely, we show that the success rate increases when traces of higher dimensionality are used. We also show that template attacks (without initial dimensionality reduction using PCA, for instance) are more efficient than monovariate attacks (such as CPA [13]) applied in the first direction of the PCA leakage reduction. Such practical derivations are, to the best of our knowledge, new, and we here provide some numerical values about the actual gain conveyed by attacks of increased dimensionality.

9.1.5 Outline

In order to present our contributions, we follow the scientific approach described here-after. In Sect. 9.2, we present a mathematical modelization of the problem. In Sect. 9.3, we provide our first contribution, namely the formalization of template attacks. In Sect. 9.4, efficient algorithms to compute template attacks are given, thanks to the coalescence principle. In Sect. 9.5, we validate our contributions on real traces taken from an AES running on an ATMega 163 smart card. In particular, we exhibit a spectral approach to speed up the computation of template attacks. Finally, in Sect. 9.6, we conclude our study.

9.2 Mathematical Modelization of the Problem and Notations

9.2.1 Side-Channel Problem

Using the same notations as above, we model the side-channel problem as follows:

$$X, k \rightarrow Z \rightarrow M(Z) = M \rightarrow L = M + B. \tag{9.1}$$

Such that, Z is an intermediate value within the targeted cryptographic algorithm, e.g., $Z = \texttt{SubBytes}(X \oplus k)$ in AES.

We assume all those variables are measured many times (N times). Therefore, the random variables have a dimensionality N, with the particularity that k is the same for all N, and that noise random variables B_q are all i.i.d.

In addition, we assume that the measurements are multidimensional of dimensionality D. This can represent the fact that

- oscilloscopes capture windows of D samples, typically at high sampling rates, resulting in traces of many samples per clock period,
- simultaneous captures by various probes (e.g., power and electromagnetic, or also multiple electromagnetic probes placed at various locations).

9.2.2 Additional Notations

We adopt the following additional notations:

- S: the cardinality of the space where Z belongs to, that is, if $Z \in \{0, 1\}^n$, then $S = 2^n$; to ease notations, we also assume (without loss of generality though) that the number of X values is S. For instance, Z typically arises from a bijective substitution box applied on $X \oplus k$.
- Σ: the $D \times D$ covariance matrix of the noise, such that $B_q \sim \mathcal{N}(0, \Sigma)$ $(0 \leq q < N)$.

Moreover, we will use the the coalescence principle. which consists in trading $\sum_q$ for $\sum_x \sum_{q:x_q=x}$, where $\sum_{q:x_q=x}$ is a shortcut notation for $\sum_{q \in \{1,...,N\}, \text{such that: } x_q=x}$.
Therefore, we have the following "types" for variables:

- X and k are the bitvectors (of second dimension N), not necessarily of the same length,
- Z is a matrix of Booleans $\{0, 1\}^n \times N$,
- M and L are the matrices of $D \times N$ real numbers.

The peculiarity of M is that we have $M(Z_q) = M(Z_{q'})$ if $X_q = X_{q'}$ (under the same key k).

We introduce the notion of *coalesced matrices*. Recall that this notion has been introduced in [14, Sect. 3.3] but was not named there. Therefore, we qualify this notion (averaging traces corresponding to the same sensitive variable) as "coalescence". It arises from the fact that M and L, random variables in Eq. (9.1), depend only on values from Z (except L which also depends on the noise, which is independent of other random variables), and therefore their matrices can be coalesced from size $D \times N$ to $D \times S$, according to the coalescence principle which is introduced and argued in Chap. 7.

Definition 9.1 (*Coalesced matrices*) Coalesced model $\tilde{M}(k)$ is the $D \times S$ matrix

$$\left(\tilde{M}_{x=0,k}, \ldots, \tilde{M}_{x=S-1,k}\right),$$

i.e., where x is enumerated in lexicographical order. This matrix is a property of the leakage from the device under test, i.e., it is not a random variable, but a constant value. For any measurement q, we denote $M_q = \tilde{M}_{x_q,k}$.

Coalesced measurement $\tilde{L}$ is the matrix

$$\left(\tilde{L}_0, \ldots \tilde{L}_{S-1}\right) = \left(\frac{\sum_{q:x_q=0} L_q}{\sum_{q:x_q=0} 1}, \ldots, \frac{\sum_{q:x_q=S-1} L_q}{\sum_{q:x_q=S-1} 1}\right).$$

In this equation, we denote by n_x the number of plaintexts such that x_q equals to x. If for some x the number n_x it is equal to zero, then, by convention $\tilde{L}_x = 0$. Alternatively, the empty classes can be pruned. Both choices are equivalent.

The advantage of Definition 9.1 is twofold. First, as long as $N \geq \#X = 2^n$, the matrix L (of size $D \times N$) is reduced to a matrix $\tilde{L}$ (of size $D \times 2^n$). Second, when n_x is not equal to zero, $\tilde{L}_x$ as defined here is the empirical average of $\mathbb{E}(L|X = x)$ over n_x values. Thus, for any x and q, L_q (for $x_q = x$) and $\tilde{L}_x$ have same expectation. But, as stated by Cochran's Theorem [15], $\tilde{L}_x$ has an empirical standard deviation divided by $\sqrt{n_x}$ relatively to that of L_q.

It shall also be noted that, in our model, the $D \times N$ matrix M (and the $D \times S$ coalesced matrix $\tilde{M}$) are general. In particular, we do not assume any structure, such as $\tilde{M}$ being the product of a $D \times 1$ column by a $1 \times S$ row, as done in [1, 16].

Examples of Z functions are given hereafter:

- in the software case, $X, k \in \{0, 1\}^n$, and $Z = X \oplus k$;
- in the hardware case, for the case of AES ($n = 8$) attacked on the last round:

  - $X, k \in \{0, 1\}^n$, and $Z = (\text{InvSubBytes}(X \oplus k), X)$ for the bytes of the first row, and
  - $X = (X_1, X_2) \in \{0, 1\}^n \times \{0, 1\}^n$,
    and $Z = (\text{InvSubBytes}(X_1 \oplus k), X_2)$ for the other three rows.

Those two cases refer to the two models expressed in [17]:

- Only manipulated Data Leak (ODL): only the manipulated value influences the leakage.
- Memory Transitions Leak (MTL): two values (the previous one and the new one) of a memory unit (e.g., a register) influence the power consumption, and also the device leaks some combination of the two consecutively manipulated values.

We also introduce other useful notations for matrices:

- Let A be a square matrix, then its trace is the sum of elements along its diagonal $\text{tr}(A) = \sum_i A_{i,i}$; let A and A' two rectangular matrices such that AA' and $A'A$ are square. Then the trace has the property that $\text{tr}(AA') = \text{tr}(A'A)$.
- Covariance matrices are symmetrical and all their eigen-values are positive. When some eigen-values are zero, it means that rows are redundant, and those are implicitly removed until all eigen-values are strictly positive. Alternatively, more measurements shall be captured. Therefore, we will consider covariance matrices are invertible. For example, the noise covariance matrix Σ has an inverse, denoted Σ^{-1}. We will also make use of the notation $\Sigma^{-1/2}$ for the (unique[1]) matrix such

---

[1] Recall that Σ is a symmetric matrix. Therefore, there exists one invertible matrix P such that $\Sigma = PDP^{-1}$, where D is a diagonal matrix whose diagonal coefficients are all positive. It is customary to call the diagonal coefficients of D the eigen-values of Σ and P the matrix of eigen-vectors of Σ. Let us denote $D^{1/2}$ the diagonal matrix where diagonal coefficients are the square root of those of D. Then, $\Sigma^{1/2} := PD^{1/2}P^{-1}$ matches the definition, since $\Sigma^{1/2}\Sigma^{1/2} = PD^{1/2}P^{-1}PD^{1/2}P^{-1} = PD^{1/2}D^{1/2}P^{-1} = PDP^{-1} = \Sigma$.

that,
$$\Sigma^{-1/2}\Sigma^{-1/2} = \Sigma^{-1}.$$

Eventually, let us define the following operator:

Definition 9.2 (*The k-th trace operator* tr_k) Let $k \in \mathbb{F}_2^n$. The k-th trace operator tr_k of a $2^n \times 2^n$ square matrix M is

$$\mathrm{tr}_k(M) = \sum_x M_{x,x\oplus k}.$$

The regular trace of a square matrix M is simply $\mathrm{tr} = \mathrm{tr}_0$.

9.3   Formalization of Template Attacks

In this section, we present our main result. We formalize the two phases of the template attack, namely learning and attacking. Their algorithms reveal a very simple mathematical expression for building the template model and for matching never-seen traces (an expression which happens to be very similar). In particular, this leads us to simplify the computation for the two phases: the mathematical expressions involve only linear algebra (neither logarithms nor exponential functions need to be evaluated); only one matrix inversion is required (namely, once at the end of the profiling stage). Moreover, we show that the N traces used either for profiling or matching can be regrouped in $S = 2^n$ classes. In the sequel, we use 2^n (e.g., $2^n = 256$ for the AES) instead of S for the work to be concrete. The attack two phases can in turn be computed using 2^n-dimensional vectors and matrices of dimension $2^n \times 2^n$. We will say that the number of traces N is coalesced into 2^n classes, which further simplifies the computations.

We start by analyzing the classical template attacks (without coalescence) in Sect. 9.3.1. We then see the gain of considering template attacks with coalesced traces and models in Sect. 9.3.2. Eventually, we compare results with attacks making use of dimensionality reduction in Sect. 9.3.3.

9.3.1   Template Attack (Without Coalescence)

Template Building = Profiling Stage
In the profiling stage, the attacker computes $\tilde{M}$, a $D \times 2^n$ matrix, and Σ. An experimental method to profile is the following:

- In the estimation of $\tilde{M}$, the attacker fixes Z to some all possible values (the column index of $\tilde{M}$), and averages many measurements, so as cancel out the noise; in order to explore all the possible values of Z, the attacker can enumerate all values of X

and k. However, due to symmetries, it might be possible that some values of Z be encountered by many different pairs (X, k), hence there is an opportunity to save some profiling effort.

- Regarding the estimation of Σ, the attacker fixes Z to an arbitrary value, say 0, and then estimates Σ as the covariance matrix of the traces.

Therefore, in the sequel, we consider that the averages of the traces ($\tilde{M}$) and the noise covariance matrix (Σ) are known. Notice that $B \sim \mathcal{N}(0, \Sigma)$ has zero mean because the data (data = text and key)-dependent part of the traces constitutes the expectation of the noise.

Also notice that we implicitly opted for a leakage decomposition basis known as the *canonical basis* [18]. However, any other choice of bases suits perfectly, insofar as the change to another basis is a multiplication by an invertible $2^n \times 2^n$ matrix, which can be applied on coalesced matrices (the noise is not impacted by the basis change, which concerns only the data). Such basis change could be interesting if the leakage happens to be concentrated on a smaller basis than the canonical one: the uninteresting dimensions can thus safely be dropped, thereby simplifying the equations.

Template Attack

After the *identification* stage (profiling, as described in Sect. 9.3.1), the attacker can perform the *exploitation* stage. We have the following result:

Theorem 9.1 (Theorem 1 of [19]) *Template attacks guess the key as follows:*

$$\hat{k} = \underset{k}{\mathrm{argmin}}\, \mathrm{tr}((L - M_k)^\mathsf{T} \Sigma^{-1}(L - M_k)). \tag{9.2}$$

Standardization of Traces

Theorem 9.2 (Template attack with standardized traces) *It is possible to standardize the templates (and the traces), by trading*

- M_k *for* $M'_k = \Sigma^{-1/2} M_k$*, and*
- L *for* $L' = \Sigma^{-1/2} L$.

Accordingly, the template attack simplifies as follows:

$$\hat{k} = \underset{k}{\mathrm{argmin}}\, \mathrm{tr}((L' - M'_k)^\mathsf{T}(L' - M'_k)) = \underset{k}{\mathrm{argmin}}\, ||L' - M'_k||_F^2, \tag{9.3}$$

where $|| \cdot ||_F^2$ is the square Frobenius norm of a matrix, that is the sum of all its elements raised to the power two.

Proof Notice that, owing to the symmetry of Σ, one also has the following property $\left(\Sigma^{-1/2}\right)^{\mathsf{T}} = \Sigma^{-1/2}$. Hence,

$$
\begin{aligned}
(L - M_k)^{\mathsf{T}}\Sigma^{-1}(L - M_k) &= (L - M_k)^{\mathsf{T}}\Sigma^{-1/2}\Sigma^{-1/2}(L - M_k) \\
&= (\Sigma^{-1/2}(L - M_k))^{\mathsf{T}}\Sigma^{-1/2}(L - M_k) \\
&= (L' - M_k')^{\mathsf{T}}(L' - M_k').
\end{aligned}
$$

Besides,

$$
\begin{aligned}
\mathrm{tr}&((L' - M_k')^{\mathsf{T}}(L' - M_k')) \\
&= \sum_q (L_q' - M_{q,k}')^{\mathsf{T}}(L_q' - M_{q,k}') \\
&= \sum_q \|L_q' - M_{q,k}'\|_2^2 \qquad \text{(norm-2 of a D-dimensional vector)} \\
&= \|L' - M_k'\|_F^2 \qquad \text{(Frobenius norm of a $D \times N$ matrix).}
\end{aligned}
$$

$\qquad\square$

Notice that the standardized noise $B' = \Sigma^{-1/2}B$ has distribution $\mathcal{N}(0, I)$, where I is the $D \times D$ identity matrix.

Remark 9.1 In the expression of the template attack of Eq. (9.3), the covariance matrix Σ disappears: it is hidden half in the model (M' is $\Sigma^{-1/2}M$) and half in the matching traces (L' is $\Sigma^{-1/2}L$). Alternatively, we will show in Algorithm 9.8 that Σ can be completely hidden in the templates, and that there is no need to use Σ in the corresponding matching phase (Algorithm 9.9). This way of using Σ allows to minimize the computation time, in particular, because Σ will need to be inverted only once, namely when building the model.

9.3.2 Template Attack (With Coalescence)

The trace operator in Theorem 9.1 is applied to a $N \times N$ matrix (resulted from the product of the raw matrices, without coalescence), which poses a problem of scaling as the number of traces grows. Therefore, we show that template attacks can be rewritten in a coalesced form:

Proposition 9.1 (Coalesced template attack) *Template attacks (as per (9.2)) guess the key as follows:*

$$
\hat{k} = \underset{k}{\arg\min} \sum_x n_x (\tilde{L}_x - \tilde{M}_{x,k})^{\mathsf{T}}\Sigma^{-1}(\tilde{L}_x - \tilde{M}_{x,k}), \tag{9.4}
$$

where $n_x = \sum_{q:x_q=x} 1$ is the number of traces corresponding to plaintext value x.

Proof By developing the argument to minimize in Theorem 9.1, there remain only two terms which depend on the key k:

$$\sum_q M_{x_q,k}^{\mathsf{T}} \Sigma^{-1} L_q = L_q^{\mathsf{T}} \Sigma^{-1} M_{x_q,k} \tag{9.5}$$

and

$$\sum_q M_{x_q,k}^{\mathsf{T}} \Sigma^{-1} M_{x_q,k}, \tag{9.6}$$

because $\sum_q L_q^{\mathsf{T}} \Sigma^{-1} L_q$ is independent from the key k.

Now,

$$\sum_q M_{x_q,k}^{\mathsf{T}} \Sigma^{-1} L_q = \sum_x \sum_{q:x_q=x} M_{x_q,k}^{\mathsf{T}} \Sigma^{-1} L_q \qquad = \sum_x \left(\sum_{q:x_q=x} M_{x_q,k}^{\mathsf{T}} \Sigma^{-1} L_q \right)$$

$$= \sum_x \left(\sum_{q:x_q=x} \tilde{M}_{x,k}^{\mathsf{T}} \Sigma^{-1} L_q \right) \qquad = \sum_x \tilde{M}_{x,k}^{\mathsf{T}} \Sigma^{-1} \left(\sum_{q:x_q=x} L_q \right)$$

$$= \sum_x n_x \tilde{M}_{x,k}^{\mathsf{T}} \Sigma^{-1} \left(\frac{\sum_{q:x_q=x} L_q}{n_x} \right) = \sum_x n_x \tilde{M}_{x,k}^{\mathsf{T}} \Sigma^{-1} \tilde{L}_x$$

and

$$\sum_q M_{x_q,k}^{\mathsf{T}} \Sigma^{-1} M_{x_q,k} = \sum_x \sum_{q:x_q=x} M_{x_q,k}^{\mathsf{T}} \Sigma^{-1} M_{x_q,k} \qquad = \sum_x \sum_{q:x_q=x} \tilde{M}_{x,k}^{\mathsf{T}} \Sigma^{-1} \tilde{M}_{x,k}$$

$$= \sum_x \tilde{M}_{x,k}^{\mathsf{T}} \Sigma^{-1} \tilde{M}_{x,k} \left(\sum_{q:x_q=x} 1 \right) = \sum_x n_x \tilde{M}_{x,k}^{\mathsf{T}} \Sigma^{-1} \tilde{M}_{x,k}.$$

All in one, up to a key-independent term $\sum_x n_x \tilde{L}_x^{\mathsf{T}} \Sigma^{-1} \tilde{L}_x$, we can rewrite the argument to minimize in Theorem 9.1 as:

$$\sum_x n_x (\tilde{L}_x - \tilde{M}_{x,k})^{\mathsf{T}} \Sigma^{-1} (\tilde{L}_x - \tilde{M}_{x,k}).$$

$\square$

Remark 9.2 The equation (9.4) reads as a trace on the plaintext space, weighted by the values n_x. Regarding n_x, they are the empirical estimators of $N\mathbb{P}(X = x)$.

Remark 9.3 The attack presented in Proposition 9.1 is exactly the same as that of Theorem 9.1: it will succeed with the same number of traces. However, the attack in Proposition 9.1 is computationally more efficient than that of Theorem 9.1 as soon as the number of traces N is larger than the number of plaintexts involved in leakage model 2^n (for AES $2^n = 256$). The gain holds both for training and matching phases. However, we underline that training requires much more traces than matching, hence most of the gain of using coalesced data arises from the template building phase.

9.3.3 State-of-the-Art Dimensionality Reduction for TA

It has been suggested in [5] that template attacks can be made more practical by

1. first start with a dimensionality reduction,
2. then perform a template attack in a reduced space.

This approach deserves to be confronted with the *innate* dimensionality reduction power of template attacks, as expressed in Sect. 9.3.2 (where the complexity is related to cryptographic parameter $S = 2^n$ and not to the traces dimensionality D).

To be more accurate, two-dimensionality reduction techniques have been put forward: PCA [5] and Linear Discriminant Analysis (LDA [20]).

In this section, we contrast template attacks with and without such pre-processing (especially the PCA).

PCA

The PCA is a technique aiming at identifying linear projections which concentrate information in few directions, namely the principal directions. In this respect, it is already pointed out in [5] that any set of training traces of high dimensionality D can always be reduced to lower dimensionality 2^n, as we have also noted in Sect. 9.3.2.

The PCA appears in two variants: *classical* and *class-based*, as tailored by Archambeau et al. in [5]. Both techniques require the definition of a *covariance matrix*. We analyze in the following lemma both techniques, which resort to the concept of *coalesced matrices* (Definition 9.1):

Definition 9.3 (*Variance of a vectorial random variable*) The variance of $L \in \mathbb{R}^D$ is $\mathrm{VAR}(L) = \mathbb{E}((L - \mathbb{E}(L))(L - \mathbb{E}(L))^{\mathsf{T}})$.

Lemma 9.1 (Law of total variance)
$$\mathrm{VAR}(L) = \mathbb{E}(\mathrm{VAR}(L|X)) + \mathrm{VAR}(\mathbb{E}(L|X)).$$

Proof Adaptation of the law of total variance to the multivariate case. □

Applied to $L = M + B$, where M depends on X (but B does not), we have
$$\mathrm{VAR}(L) = \mathrm{VAR}(M) + \Sigma.$$
Besides, as $M|X$ is deterministic, we can apply Lemma 9.1
$$\mathrm{VAR}(M) = \mathrm{VAR}(\mathbb{E}(M|X)).$$
Let us assume M is centered, namely $\mathbb{E}(M) = 0$. Hence,
$$\mathrm{VAR}(M) = \sum_x \mathbb{P}(X = x)\mathbb{E}(M|X)\mathbb{E}(M|X)^{\mathsf{T}}.$$

Lemma 9.2 (PCA) *We resort to the Law of Large Numbers (LLN). The* classical *PCA is based on the estimation of the following $D \times D$ covariance matrix:*

$$\frac{1}{N} \sum_{q=1}^{N} \left(L_q - \frac{1}{N} \sum_{q'} L_{q'} \right) \left(L_q - \frac{1}{N} \sum_{q'} L_{q'} \right)^{\mathsf{T}} \xrightarrow[N \to +\infty]{LLN} \mathrm{VAR}(M) + \Sigma. \quad (9.7)$$

The PCA of Archambeau et al. is based on the estimation of the following $D \times D$ covariance matrix:

$$\frac{1}{N} \sum_{q=1}^{N} \left(\tilde{L}_q - \frac{1}{N} \sum_{q'} \tilde{L}_{q'} \right) \left(\tilde{L}_q - \frac{1}{N} \sum_{q'} \tilde{L}_{q'} \right)^{\mathsf{T}} \xrightarrow[N \to +\infty]{LLN} \mathrm{VAR}(M). \quad (9.8)$$

Regarding the PCA, we only consider the second case. As the traces are first averaged, the covariance arising from the noise disappears; therefore, the only contribution is the intra-class variability. As Σ is a covariance matrix, it has only positive eigen-values. We consider the set of corresponding eigen-vectors as a matrix V (matrix of eigen-vectors), which is such that

$$\tilde{M}_k (\tilde{M}_k)^{\mathsf{T}} V = V \Delta,$$

where the matrix $\Delta = \mathrm{diag}(\lambda_1, \lambda_2, \ldots, \lambda_{2^n})$ is the diagonal matrix of eigen-values. Besides, it is known [21] that $V V^{\mathsf{T}} = I$, the $D \times D$ identity matrix.

Lemma 9.3 *Template attack with PCA is always less efficient (in terms of success probability) than the template attack.*

Proof The reason is simply because the maximum likelihood is the best attack in terms of success probability. As the PCA is a pre-processing, it is less efficient. □

Notice that, in practice, it is hard to estimate models and noise accurately, therefore reducing the dimensionality before attacking is a winning strategy. However, assuming that a perfect profiling is possible, any pre-processing reduces the information (recall the "data processing theorem"), therefore one may wonder how different efficiencies of template attacks *with* and *without* PCA are.

In the case $D < 2^n$, $\tilde{M}_k(\tilde{M}_k)^{\mathsf{T}}$ is almost of full rank (with very high probability). However, as traces are centered, there is one relationship amongst all rows of $\tilde{M}_k$, namely their sum is null. Hence, V is not invertible. There is actually no dimensionality reduction: the projection on any eigen-vector carries information.

The PCA therefore consists in transforming the traces the following way:

- $\tilde{M}$ becomes $V^{\mathsf{T}} \tilde{M}$ and L becomes $V^{\mathsf{T}} L$ (projection),
- the covariance matrix Σ becomes $V^{\mathsf{T}} \Sigma V$.

Table 9.1 Dimensions of traces and models, seen as matrices in this chapter

	Raw	Coalesced
Traces	$L : D \times N$	$\tilde{L} : D \times 2^n$
Models	$M : D \times N$	$\tilde{M} : D \times 2^n$

Indeed, the covariance matrix of the projected (centered) noise is $\mathbb{E}((V^{\mathsf{T}} B)$
$(V^{\mathsf{T}} B)^{\mathsf{T}}) = \mathbb{E}(V^{\mathsf{T}} B B^{\mathsf{T}} V) = V^{\mathsf{T}} \mathbb{E}(B B^{\mathsf{T}}) V = V^{\mathsf{T}} \Sigma V$.

The argument to maximize over k in the optimal distinguisher (9.2) becomes

$$\mathrm{tr}\left(\left(\left(V^{\mathsf{T}} L - V^{\mathsf{T}} M_k\right)^{\mathsf{T}} \left(V^{\mathsf{T}} \Sigma V\right)^{-1} \left(V^{\mathsf{T}} L - V^{\mathsf{T}} M_k\right)\right)\right) =$$
$$\mathrm{tr}\left(\left((L - M_k)^{\mathsf{T}} V \left(V^{\mathsf{T}} \Sigma V\right)^{-1} V^{\mathsf{T}} (L - M_k)\right)\right) =$$
$$\mathrm{tr}\left(\left((L - M_k)^{\mathsf{T}} \left(V \left(V^{\mathsf{T}} \Sigma V\right)^{-1} V^{\mathsf{T}}\right) (L - M_k)\right)\right). \tag{9.9}$$

Assume that V is invertible. Using the property that $(A A')^{-1} = A'^{-1} A^{-1}$, one has

$$V \left(V^{\mathsf{T}} \Sigma V\right)^{-1} V^{\mathsf{T}} = V V^{-1} \Sigma^{-1} (V^{\mathsf{T}})^{-1} V^{\mathsf{T}} = \Sigma^{-1},$$

which is the same covariance matrix as for the optimal distinguisher (matrix V disappears). But V is not invertible, and therefore the Eq. (9.9) cannot be simplified.

9.4 Efficiently Computing Templates with Coalescence

9.4.1 Simplification by the Law of Large Number (LLN)

In this section, we come back to genuine template attacks, as described in Sect. 9.3.

In the case of the EIS, and in the specific case when $M_{x,k}$ depends only on $x \oplus k$, we can drop the $\tilde{M}_{x,k}^{\mathsf{T}} \Sigma^{-1} \tilde{M}_{x,k}$ term. The effect is represented in Fig. 9.1: notwithstanding that the value of the distinguisher (Eq. (9.5)) is always the largest for the correct key $k = k^*$, one can see "oscillations" occurring every $2^n = 16$ traces (here $n = 4$), where indeed the dropped (Eq. (9.6)) term is the same for all key hypotheses. We recall the dimension of matrices in Table 9.1.

The maximum likelihood attack minimizes over k

- either: $\mathrm{tr}\left((L - M_k)^{\mathsf{T}} \Sigma^{-1} (L - M_k)\right)$
- or equivalently $\sum_{x \in \mathbb{F}_2^n} n_x (\tilde{L}_x - \tilde{M}_{x,k})^{\mathsf{T}} \Sigma^{-1} (\tilde{L}_x - \tilde{M}_{x,k})$.

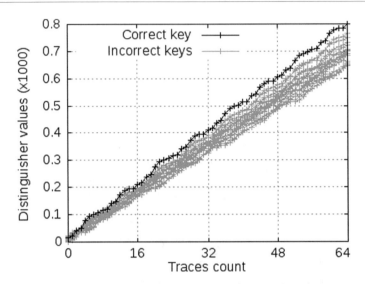

Fig. 9.1 Values of the distinguisher (Eq. (9.5)), where the $\tilde{M}_{x,k}^{\mathsf{T}}\Sigma^{-1}\tilde{M}_{x,k}$ term (Eq. (9.6)) is dropped

Using the law of large numbers, for large N, we have

$$\frac{1}{N}\sum_{x\in\mathbb{F}_2^n} n_x(\tilde{M}_{x,k})^{\mathsf{T}}\Sigma^{-1}\tilde{M}_{x,k} \xrightarrow[N\to+\infty]{} \sum_x \mathbb{P}(x)(\tilde{M}_{x\oplus k})^{\mathsf{T}}\Sigma^{-1}\tilde{M}_{x\oplus k}$$

which does not depend on k if X is uniformly distributed, that is $\mathbb{P}(x) = 2^{-n}$. This is a mathematical justification for the observation made in Fig. 9.1. As already noted in Sect. 9.3.2, there is therefore only one key-dependent term which remains

$$\sum_x n_x\tilde{L}_x^{\mathsf{T}}\Sigma^{-1}\tilde{M}_{x\oplus k} = \mathsf{tr}_k(\tilde{L}^{\mathsf{T}}\Sigma^{-1}\tilde{M}), \tag{9.10}$$

where we recall that tr_k is the k-th trace operator introduced in Definition 9.2.

The templates are estimated as represented in Algorithm 9.6. In this formulation, it clearly appears that there is a unique noise covariance matrix Σ, which justifies the technique of "pooled covariance matrix" presented in [4]. Indeed, by the law of large numbers, we have that, for all $x \in \mathbb{F}_2^n$

$$\lim_{N\to+\infty}\frac{1}{N}\sum_{q:x_q=x} L_q = \lim_{N\to+\infty}\frac{1}{N}\sum_{q:x_q=x}\tilde{M}_{x_q}$$

$$= \lim_{N\to+\infty}\frac{1}{N}\sum_{q:x_q=x}\tilde{M}_x$$

$$= \lim_{N\to+\infty}\frac{\tilde{M}_x}{N}\sum_{q:x_q=x} 1$$

$$= \lim_{N\to+\infty}\frac{n_x}{N}\tilde{M}_x \approx \frac{1}{2^n}\tilde{M}_x,$$

since $n_x \approx N/2^n$ when N is large. Besides, as $L_q = \tilde{M}_{x_q} + B$, we have that

$$\Sigma = \frac{1}{N} \sum_{q=1}^{N} (L_q - \tilde{M}_{x_q})(L_q - \tilde{M}_{x_q})^{\mathsf{T}}$$

$$= \frac{1}{N} \sum_{q=1}^{N} L_q L_q^{\mathsf{T}}$$

$$- \frac{1}{N} \sum_{q=1}^{N} L_q \tilde{M}_{x_q}^{\mathsf{T}} - \frac{1}{N} \sum_{q=1}^{N} \tilde{M}_{x_q} L_q^{\mathsf{T}} + \frac{1}{N} \sum_{q=1}^{N} \tilde{M}_{x_q} \tilde{M}_{x_q}^{\mathsf{T}}$$

$$\xrightarrow[N \to +\infty]{} \frac{1}{N} LL^{\mathsf{T}} - \frac{1}{2^n} \tilde{M} \tilde{M}^{\mathsf{T}},$$

assuming $n_x/N \to 1/2^n$ when $N \to +\infty$ (plaintext uniformity). This justifies line 9 of Algorithm 9.6.

9.4.2 Profiling and Attack Algorithms

The correct key is recovered using Algorithm 9.7. In this algorithm, the matrices $\tilde{L}$ and $\tilde{M}$ are respectively $\tilde{L} = (\tilde{l}_0, \cdots, \tilde{l}_{2^n-1})$ and $\tilde{M} = (\tilde{m}_0, \cdots, \tilde{m}_{2^n-1})$. Premise of formal presentation of template attacks, including the notion of Mahalanobis measure is presented in a paper by Zhang et al. [22].

When the templates $(\tilde{M}, \Sigma)$ are obtained from Algorithm 9.6, then the attack is no longer optimal [23], but is termed template attack. However, template attacks tend to optimal attack when the profiling is carried out on a number of traces which tends to infinity.

9.4.3 Improved Profiling and Attack Algorithms

In practice, the profiling (Algorithm 9.6) should be carried out only once, whereas the matching (Algorithm 9.7) should be carried out every time when the attack outcome is needed. So, the adversary can compute $\Sigma^{-1}\tilde{M}$ once t the end of the first stage (Algorithm 9.8 instead of Algorithm 9.6), and he can use this result every time the attack needs to be estimated (Algorithm 9.9 instead of Algorithm 9.7). Finally, it is possible to regroup the accumulations (lines 1–8) of Algorithms 9.6, 9.7, 9.8, and 9.9 in the same function.

9.4.4 Extension of Our Approach to Masked Implementations

Let d be a strictly positive integer. The protection by a masking of order d consists in dividing randomly each sensitive variable Z into at least $(d + 1)$ shares $Z^{(0)}$, ..., $Z^{(d)}$. In order to reveal this secret value, an adversary must carry out a so-called

Algorithm 9.1 Estimation of templates.

input   : Set of *training* traces L and corresponding plaintexts X for the estimation of the templates, for a known key k

output   : Templates:
- $\tilde{M}$, the matrix of the templates, and
- Σ, its homoscedastic covariance matrix.

1 for $x \in \mathbb{F}_2^n$ **do**	// Initialization
2 $\quad$ $\tilde{m}_x \leftarrow 0$	// Mean trace per class
3 $\quad$ $n_x \leftarrow 0$	// Number of traces per class
4 for $q \in \{1, \ldots, N\}$ **do**	// Accumulation, as in [14]
5 $\quad$ $\tilde{m}_{x_q \oplus k} \leftarrow \tilde{m}_{x_q \oplus k} + L_q$	
6 $\quad$ $n_{x_q \oplus k} \leftarrow n_{x_q \oplus k} + 1$	
7 for $x \in \mathbb{F}_2^n$ **do**	// Normalization
8 $\quad$ $\tilde{m}_x \leftarrow \tilde{m}_x / n_x$	

9 return $\left(\tilde{M} = (\tilde{m}_0, \cdots, \tilde{m}_{2^n-1}), \Sigma = \frac{1}{N} L L^\mathsf{T} - \frac{1}{2^n} \tilde{M} \tilde{M}^\mathsf{T} \right)$

Algorithm 9.2 Key recovery using optimal distinguisher, applied on profiling generated by Alg. 9.1.

input   : Set of *matching* traces L, corresponding plaintexts X and the templates $(\tilde{M}, \Sigma)$ obtained from Alg. 9.1

output   : Key guess k recovered by the optimal distinguisher

// $\cdots\cdots\cdots\cdots\cdots\cdots\cdots\cdots\cdots\cdots\cdots\cdots\cdots\cdots\cdots\cdots\cdots$ Accumulation, as in lines 1-8 of Alg. 9.1	
1 for $x \in \mathbb{F}_2^n$ **do**	// Initialization
2 $\quad$ $\tilde{l}_x \leftarrow 0$	// Mean trace per class
3 $\quad$ $n_x \leftarrow 0$	// Number of traces per class
4 for $q \in \{1, \ldots, N\}$ **do**	// Accumulation
5 $\quad$ $\tilde{l}_{x_q} \leftarrow \tilde{l}_{x_q} + L_q$	
6 $\quad$ $n_{x_q} \leftarrow n_{x_q} + 1$	
7 for $x \in \mathbb{F}_2^n$ **do**	// Normalization
8 $\quad$ $\tilde{l}_x \leftarrow \tilde{l}_x / n_x$	
// $\cdots\cdots\cdots\cdots\cdots\cdots\cdots\cdots\cdots\cdots\cdots\cdots\cdots\cdots\cdots\cdots$ Attack proper	

9 return $\underset{k}{\operatorname{argmax}} \operatorname{tr}_k \left(\tilde{L}^\mathsf{T} \Sigma^{-1} \tilde{M} \right)$ $\qquad$ // Eqn. (9.10)

---

high-order SCA. It consists in combining the leakage corresponding to all the $(d+1)$ shares. For such high-order attack, we leverage same notations as in Sect. 9.2, with the following extensions:

- $(L^{(w)})_{0 \leq w \leq d}$ denote the measured leakage corresponding, respectively, to the $(d+1)$ shares;
- $(M^{(w)})_{0 \leq w \leq d}$ denote the leakage model corresponding, respectively, to the $(d+1)$ shares.

Algorithm 9.3 Estimation of templates, improved by caching matrix Σ inversion at the end.

---

input : Set of *training* traces L and corresponding plaintexts X for the estimation of the templates, for a known key k

output : The matrix of templates left multiplied by the inverse of the covariance matrix $\tilde{\mathbb{M}} = \Sigma^{-1} \tilde{M}$

---

1 **for** $x \in \mathbb{F}_2^n$ **do**   // Initialization
2     $\tilde{m}_x \leftarrow 0$ // Mean trace per class
3     $n_x \leftarrow 0$ // Number of traces per class

4 **for** $q \in \{1, \ldots, N\}$ **do**   // Accumulation, as in spsciteDBLP:confspsasiacryptspsLomnePR13
5     $\tilde{m}_{x_q \oplus k} \leftarrow \tilde{m}_{x_q \oplus k} + L_q$
6     $n_{x_q \oplus k} \leftarrow n_{x_q \oplus k} + 1$

7 **for** $x \in \mathbb{F}_2^n$ **do**   // Normalization
8     $\tilde{m}_x \leftarrow \tilde{m}_x / n_x$

9 $\tilde{M} \leftarrow (\tilde{m}_0, \cdots, \tilde{m}_{2^n - 1})$ // Same as in Alg. 9.1
10 $\Sigma \leftarrow \frac{1}{N} L L^\mathsf{T} - \frac{1}{2^n} \tilde{M} \tilde{M}^\mathsf{T}$ // Same as in Alg. 9.1
11 $\tilde{\mathbb{M}} \leftarrow \Sigma^{-1} \tilde{M}$ // Costly operation, factored here for all subsequent uses in Alg. 9.4
12 **return** $\tilde{\mathbb{M}}$

To extend our template approach to masking-based protected implementations, we take as a starting point the *normalized* product combination [24]. In fact, one can see the measured trace as a set of $(d + 1)$ sub-traces, such that each sub-trace corresponds to one of the $(d + 1)$ shares. According to [24], an adversary which combines $(d + 1)$ relevant PoIs (one from each sub-trace) should be successful in his attack by carrying out a *normalized* product combination accompanied with $\mathbb{E}\left(\Pi_{w=0}^d (M^{(w)} - \mathbb{E}(M^{(w)})) \mid Z\right)$ as an optimal model. In fact, for combining, the adversary can choose any relevant PoI of the first share, with any relevant PoI of the second share... with any relevant PoI of the $(d + 1)$th share.

Following this idea, to extend the template attack on a masked implementation, one can construct an artificial trace, such that any sample in the artificial trace is a *normalized* product combination of $(d + 1)$ PoIs (one from each sub-trace). This is illustrated in Fig. 9.2a. In this figure, only one PoI is selected from the first sub-trace, and it is combined to samples of the other d sub-traces, using the evaluator's expertise to get the best choice.

A simple way of combining is to take the same number of PoIs from each sub-trace and combine them in the same order (i.e., the ith sample, for $1 \leq i \leq D$, of the artificial trace is the combination of all the PoI of the sub-traces that are in the ith position). Fig. 9.2b describes the combination to create the artificial trace by such concatenation of D normalized products between $(d + 1)$ sub-traces. Finally, the adversary can profile the leakage model of this artificial leakage and carry out the template attack by matching the combination of current leakage (also by the normalized product) to the profiled model.

Thanks to this combination idea, one can carry out the template attack against high-order masking schemes, in a general way or following our particular approach

Algorithm 9.4 Improved key recovery using optimal distinguisher, applied on profiling as obtained from Alg. 9.3.

input	: Set of *matching* traces L, corresponding plaintexts X and the templates $(\widetilde{\mathbb{M}} = \Sigma^{-1}\widetilde{M})$
	obtained from Alg. 9.3
output	: Key guess k recovered by the optimal distinguisher

// ....Accumulation, as in lines 1-8 of Alg. 9.3

1 **for** $x \in \mathbb{F}_2^n$ **do**   // Initialization
2 $\quad \tilde{l}_x \leftarrow 0$ // Mean trace per class
3 $\quad n_x \leftarrow 0$ // Number of traces per class

4 **for** $q \in \{1, \dots, N\}$ **do**   // Accumulation
5 $\quad \tilde{l}_{x_q} \leftarrow \tilde{l}_{x_q} + L_q$
6 $\quad n_{x_q} \leftarrow n_{x_q} + 1$

7 **for** $x \in \mathbb{F}_2^n$ **do**   // Normalization
8 $\quad \tilde{l}_x \leftarrow \tilde{l}_x / n_x$

// ....Attack proper

9 **return** $\underset{k}{\operatorname{argmax}} \operatorname{tr}_k \left(\tilde{L}^\mathsf{T} \widetilde{\mathbb{M}} \right)$ // Eqn. (9.10)

(a) General case:

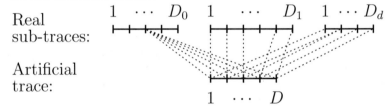

(b) Same window case:

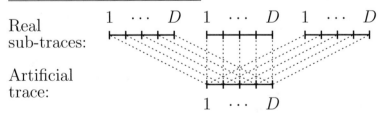

Fig. 9.2 Different methods to combine samples from $(d+1)$ sub-traces, in the case of dth-order masking countermeasure. Attack is carried out on the multivariate combination shown in the artificial trace. Each sample amongst the D making up the artificial trace is a centered product of $(d+1)$ samples taken each from a different sub-trace

(coalescence followed by the spectral approach in order to be significantly more efficient both in terms of processing time and of memory space complexities). One can see the experimental results of this technique, against a first-order Boolean masking [25], in the Sect. 9.5.5 (at scenarios 3 and 4).

9.4.5 Computational Performance Analysis

Straightforward Complexity Analysis
Let us comment on the complexity of the algorithms. The body of Algorithm 9.6, that is lines 1–8, operates on vectors of size D. The same remark applies to the body, that is lines 1–8, of Algorithm 9.7. Hence, the overall complexity of these parts of the algorithm is $D \times N$ additions.

The complex part of Algorithm 9.6, namely line 9, is computed only once. The overall complexity of this part equals to that of LL^T computation. So, that equals to $D^2 \times N$ multiplications.

The complex part of Algorithm 9.7, namely line 9, is also computed only once. Firstly, since the profiling stage should be done only one time, we can compute $\Sigma^{-1}\tilde{M}$ once, as shown in Algorithm 9.8. The overall complexity of this multiplication equals $D^2 \times 2^n$. For inverting Σ efficiently, one can use the optimized Coppersmith-Winograd-like algorithm, that has a complexity of $\mathcal{O}(d^{2.373})$ [26].

Another advantage of our template analysis is that the overall complexity of the attack phase, namely Algorithm 9.9, does not depend on N, no matter how large N is. Indeed, the overall complexity of $\tilde{L}^T(\Sigma^{-1}\tilde{M})$ computation is equal to $2^{2n} \times D$ multiplications. Table 9.2 shows the computing time of Algorithm 9.8 according to D. It can be seen that compute matrix produces to derive Σ takes more time than the inverse of Σ. Since the time computation of Algorithm 9.8 depends only on lines 10 and 11 (asymptotically), we provide its duration in Table 9.2. Of course, the overall time of Algorithm 9.8 is about the sum of lines 10 and 11.

In order to save memory space, we compute Σ in line 10 of Algorithm 9.8 from L and $\tilde{M}$ without using any temporary matrix. Finally, we can factor the code of lines 1–8 of Algorithm 9.6, 9.7, 9.8, and 9.9 in the same function.

Table 9.2 Computing time (in seconds) of Algorithm 9.8 according to D

	100	200	300	400	500	600	700	800	900	1000
Line 10 of Algorithm 9.8	0.252	0.869	1.904	3.354	5.197	7.488	10.147	13.259	16.769	20.740
Line 11 of Algorithm 9.8	0.031	0.177	0.528	1.155	2.103	3.486	5.362	7.794	10.903	16.021
Inversion (Σ^{-1})	0.012	0.093	0.314	0.751	1.472	2.575	4.133	6.183	8.814	12.112
Product ($\Sigma^{-1}\tilde{M}$)	0.019	0.084	0.214	0.404	0.631	0.911	1.229	1.611	2.089	3.909

Complexity Improvement with Spectral Analysis

We improve the complexity of the template attack using the spectral approach. To compute $\text{tr}_k\left(\tilde{L}^{\mathsf{T}}\left(\Sigma^{-1}\tilde{M}\right)\right)$ more efficiently, let us first denote $\left(\Sigma^{-1}\tilde{M}\right)$ as $\tilde{\mathbb{M}}$, such that:

$$\text{tr}_k\left(\tilde{L}^{\mathsf{T}}\left(\Sigma^{-1}\tilde{M}\right)\right) = \text{tr}_k\left(\tilde{L}^{\mathsf{T}}\tilde{\mathbb{M}}\right).$$

Recall the dimension of $\tilde{L}^{\mathsf{T}}$ is $2^n \times D$ and that of $\tilde{\mathbb{M}}$ is $D \times 2^n$. The (i, j) element of the matrix $\tilde{L}^{\mathsf{T}}\tilde{\mathbb{M}}$ is thus

$$\tilde{L}^{\mathsf{T}}\tilde{\mathbb{M}}[i][j] = \sum_{u=0}^{D-1} \tilde{L}^{\mathsf{T}}[i][u]\tilde{\mathbb{M}}[u][j].$$

Consequently,

$$\begin{aligned}
\text{tr}_k\left(\tilde{L}^{\mathsf{T}}\tilde{\mathbb{M}}\right) &= \sum_{i=0}^{2^n-1}\sum_{u=0}^{D-1} \tilde{L}^{\mathsf{T}}[i][u]\tilde{\mathbb{M}}[u][i \oplus k] \\
&= \sum_{u=0}^{D-1}\sum_{i=0}^{2^n-1} \tilde{L}^{\mathsf{T}}[i][u]\tilde{\mathbb{M}}[u][i \oplus k] \\
&= \sum_{u=0}^{D-1} \tilde{L}^{\mathsf{T}}[.][u] \otimes \tilde{\mathbb{M}}[u][.](k) \\
&= \sum_{u=0}^{D-1} WHT^{-1}\left(WHT\left(\tilde{L}^{\mathsf{T}}[.][u]\right) \bullet WHT\left(\tilde{\mathbb{M}}[u][.]\right)\right)(k)
\end{aligned}$$

So,

$$\text{tr}_k\left(\tilde{L}^{\mathsf{T}}\tilde{\mathbb{M}}\right) = WHT^{-1}\left(\sum_{u=0}^{D-1} WHT\left(\tilde{L}^{\mathsf{T}}[.][u]\right) \bullet WHT\left(\tilde{\mathbb{M}}[u][.]\right)\right)(k), \quad (9.11)$$

where:

- $\tilde{L}^{\mathsf{T}}[.][u]$ denotes the uth column of the matrix $\tilde{L}^{\mathsf{T}}$,
- $\tilde{\mathbb{M}}[u][.]$ denotes the uth line of the matrix $\tilde{\mathbb{M}}$.

Thanks to the (normalized) Walsh-Hadamard Transform (WHT) that allows us to compute, for all $u = 0, \ldots, D-1$, the convolution product $\tilde{L}^{\mathsf{T}}[.][u] \otimes \tilde{\mathbb{M}}[u][.]$ with a complexity of $n2^n$ instead of 2^{2n}. Thereby the overall complexity of $\text{tr}_k\left(\tilde{L}^{\mathsf{T}}\left(\Sigma^{-1}\tilde{M}\right)\right)$ computation becomes $n2^n \times D$ instead of $2^{2n} \times D$. In applications such as key extraction from the AES ($n = 8$), the computation time of the attack phase is indeed divided by $2^n/n = 32$, which is a significant gain. Table 9.3 shows the computing time of the line 9 of the Algorithm 9.9 according to D, with and without the spectral analysis (resp. Eqs. 9.11 and 9.10).

Is noteworthy that this result that assumes the group operation $\oplus$ over the set $\{0, 1\}^n$ (i.e., $Z = F(x \oplus k)$) holds true for any other group operation $\odot$ as long

Table 9.3 Computing time (in seconds) of the line 9 of Algorithm 9.9 according to D, with and without the spectral analysis (WHT)

	100	200	300	400	500	600	700	800	900	1000
Without WHT	0.0619	0.1242	0.2030	0.2858	0.3447	0.4126	0.4763	0.5453	0.6151	0.6854
With WHT	0.0022	0.0042	0.0065	0.0091	0.0116	0.0135	0.0154	0.0165	0.0176	0.0201
Improv. (ratio)	28.625	29.752	30.969	31.373	29.652	30.534	30.849	32.969	34.891	34.183

as WHT is replaced by Fast Fourier Transform (FFT) on this group ($\{0, 1\}^n, \odot$). For example, since in TEA (a Tiny Encryption Algorithm) and several AES candidates [27], the operation is $+$ (mod (2^n)), so cyclical Fourier transform must replace WHT, to significantly speed up the attack.

9.5   Experiments

In this section, we assess the efficiency of this template attack. We first analyze its success rate with variable numbers of samples per trace. We compare it with the success rate of the classical CPA. In the sequel, a comparison between this template attack and the monovariate CPA attack over PCA is done.

9.5.1   Traces Used for the Case Study

An ATMega 163 smart card, involving an 8-bit AVR type processor, has been programmed to process a software AES. The analysis consists of measuring the power consumption when the AES is running. The power measurements are done with a PicoScope 3204A on the first round. The sampling rate equals 256 MSamples/second.

We first characterize the traces managed in the following experiments. The average trace, SNR (Definition 2.4.3), NICV (Definition 2.4.4), and the principal direction of PCA (recall Sect. 9.3.3) are computed in all possible samples of these traces. The Fig. 9.3 shows those characterizations over $D = 700$ samples.

In the following template attacks, a set of traces is collected for template building, while another set is collected for the attack phase. As the two sets of traces are measured from the same smart card, the results are optimistic in terms of attacker potential. We analyze the success rate of the template attack according to the size of a chosen sample window.

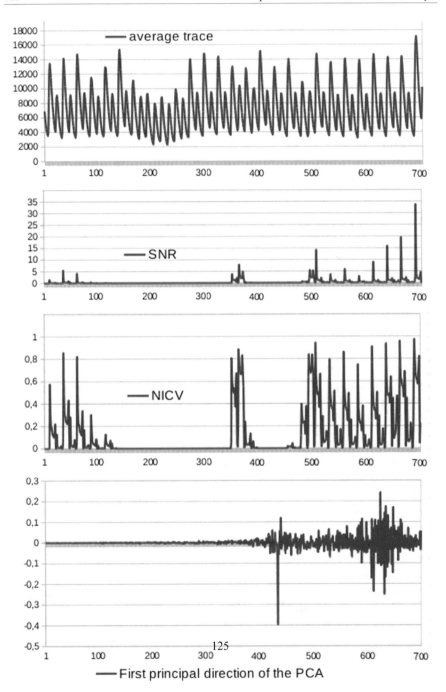

Fig. 9.3 Average trace, SNR, NICV, and first principal direction of PCA, according to the index of samples

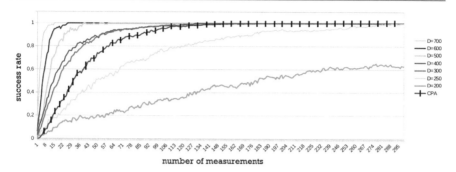

Fig. 9.4   Values of success rate, according to the number of samples

9.5.2   Template Attacks with Windows of Increasing Size

We analyzed the success rate with different numbers of samples (from $D = 1$ to $D = 700$ samples). The samples are selected around a chosen central point I_0. This point is the sample time where the classical CPA yields the highest peak. The window of growing size D gathers samples belonging to the interval $[I_0 - D/2, I_0 + D/2)$.

The Fig. 9.4 shows the success rate according to the number of measurements for different sizes of the window (from $D = 200$ to $D = 700$). From these results, one can deduce that the larger the dimensionality, the fewer the number of traces to recover the secret key. Asymptotically, for very large dimensionalities $D \rightarrow +\infty$, only one trace suffices to extract the key.

For reference, Fig. 9.4 also shows the success rate of univariate CPA (which corresponds to the maximum value of the correlation coefficient along all $D = 700$ samples). The CPA uses Hamming weight model. It can be seen that (model-agnostic) template attack becomes more efficient than CPA starting from about $D \in [250, 300]$ and keeps being better for larger dimensionality.

Indeed, as shown in Fig. 9.5, this monotony of the success rate is approximately respected above $D = 300$ samples/trace. But, as shown in Fig. 9.6, this monotony in not respected below $D = 300$. One can deduce that in this interval the success rate depends on the signal waveform; in the interval $D \in [1, 300]$, it can happen that when increasing the window size, more *noise* than *signal* is injected in the template attack, thereby making it less efficient. In order to lead this attack more efficiently, one can choose only the samples where the curve in Fig. 9.6 decreases because these samples really depend on $X \oplus k$ contrary to the samples where the curve increases.

As shown in Fig. 9.4, from around 280 samples/trace this multivariate analysis will be more efficient than the classical CPA. So one can conclude that there is a venue to capture more information using part or all of the samples rather than conducting a monovariate attack.

In order to validate our study independently from the targeted device, the same experiment is carried out on a more recent smartcard, namely TI MSP430G2553. Fig. 9.7 shows similar results on this device, comparing to the right part of Fig. 9.6 ($D \geq 300$). One can see that the more samples are taken into account, the faster

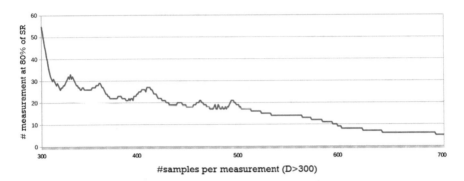

Fig. 9.5 Number of measurements at 80% of success rate, according to the number of samples (zoom for $D > 300$)

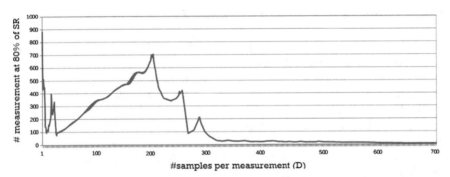

Fig. 9.6 Number of measurements at 80% of success rate, according to the number of samples (full scale, that is $D \in [1, 700]$)

the attack. This figure also shows that it is very important to choose the relevant PoIs, and that template attacks benefit from the great multiplicity of PoI. The gain is huge: more than $100\times$, less traces to recover the key when using $D = 1$ vs $D = 700$. It is noteworthy that both PoI-based and PCA-based template attacks [8] can straightforwardly benefit from our approach.

9.5.3 Comparison with PCA

In order to compare the efficiency of this method with other multivariate analyzes, the PCA analysis is carried out on the same device with the same number of samples (700 samples/trace). Fig. 9.8 shows that the template attack is more efficient in practice than both the PCA and the classical CPA. Recall that Fig. 9.3 shows the average trace, SNR, NICV, and the first principal direction of PCA, according to the index of samples.

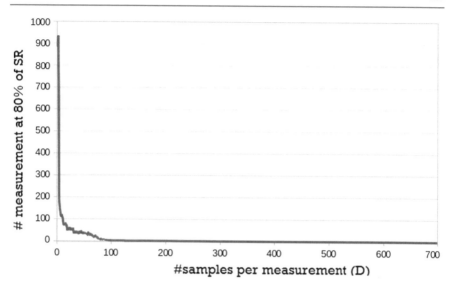

Fig. 9.7 Number of measurements at 80% of success rate, according to the number of samples (full scale, that is $D \in [1, 700]$) for the TI MSP430G2553 card

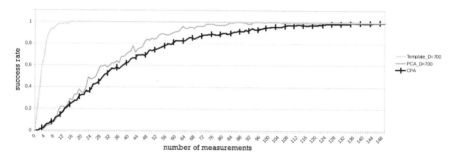

Fig. 9.8 Values of success rate, according to the number of samples

9.5.4 Template Attack after Dimensionality Reduction (over First Eigen-Components)

In this stage, we assess the effect of PCA on the success rate of our template attack. Instead of assessing the efficiency of our attack according to the number of samples per trace, it is assessed according to the number of directions of PCA ordered by decreasing eigen-value. Fig. 9.9 shows the number of measurements required to reach 80% of success rate, according to the number of eigen-components considered per measurement. This figure also shows the ordered eigen-values according to their corresponding eigen-components. It is clear that a quasi-linear relation exists between them: the success rate increases about as fast as the cumulative eigen-values.

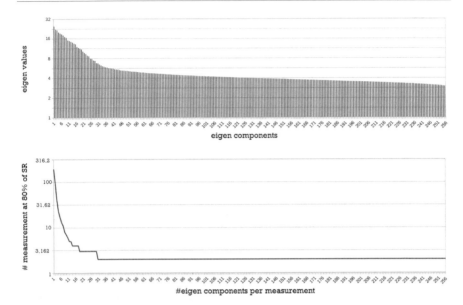

Fig. 9.9 PCA with multiple directions (from 1 to $2^n - 1$ directions). Top: scree plot of eigenvalues (log scale). Bottom: number of measurements at 80% of success rate (log scale)

9.5.5 Study of Our Approach with Simulated Traces

In order to do a fair comparison under different aspécts (PoIs, noise levels, and masking) one can resort to simulated traces. In this work, we present four scenarios with different noise levels. In each scenario, we increase a window size around a central PoI and we show the number of the required traces to reach 80% of success rate. So, we carry out, at once, a study with different PoI numbers, different noise levels, and using the masking technique or not. The four scenarios are:

1. In the first scenario, we simulate a leakage that follows a half cycle of a sine function, during 700 samples (or equivalently any 700 PoIs that follow the same law). More exactly, if the sensitive variable is Z, then the leakage at the sth time sample is

$$L_z(s) = HW(z) \sin\left(2\pi \frac{s}{2 \times 700}\right) + B,$$

such that the leakage model is $M_z(s) = HW(z) \sin\left(2\pi \frac{s}{2 \times 700}\right)$ and B denotes a Gaussian noise. In those equations, $HW(z)$ is the Hamming weight of $z \in \mathbb{F}_2^n$, that is $HW(z) = \sum_{i=1}^{n} z_i$. We used such leakage simulation in order to be close to the real leakage (e.g., see the real leakage of a smart card as shown in [28, Fig. 4]). During this experiment, we vary the window size around the central PoI ($s = I_0 = 350$). Figure 9.10 shows the number of the required traces to reach 80% of success rate according to the number of samples, for different standard deviations (σ) of the noise, in the first scenario.

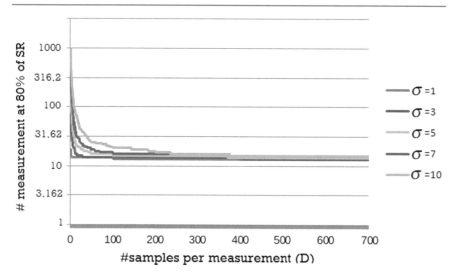

Fig. 9.10 Number of measurements at 80% of success rate (log scale), according to the number of samples, in scenario 1

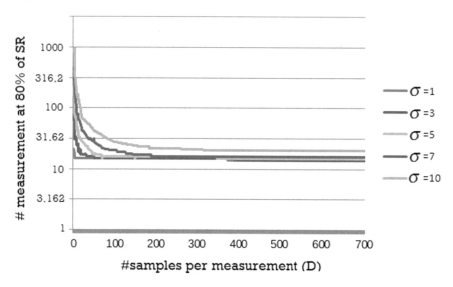

Fig. 9.11 Number of measurements at 80% of success rate (log scale), according to the number of samples, in scenario 2

2. In the second scenario, we simulate a leakage that follows the same leakage function as the first scenario but only for the even values of s. For the odd values of s, we simulate non-relevant points (non-informative points [29]) by random values. Figure 9.11 shows the number of required traces to reach 80% of success rate according to the number of samples, for different noise standard deviations

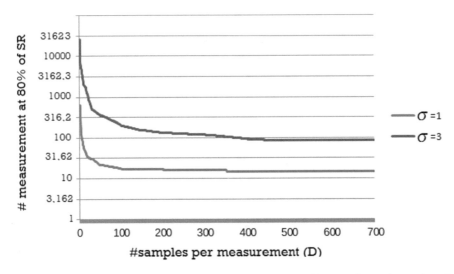

Fig. 9.12 Number of measurements at 80% of success rate (log scale), according to the number of samples, in scenario 3

(σ), in the second scenario. One can show the inconvenience of considering non-relevant points as PoIs.

3. In the third scenario, we simulate the leakage of a protected device by first-order Boolean masking [30, Sect. 4]. For each share (masked sensitive value and the mask), the leakage is simulated by a sine function as in the first scenario. Before carrying out our approach, we combine the sub-traces in one artificial trace as described in Sect. 9.4.4. Figure 9.12 shows the number of required traces to reach 80% of success rate according to the number of samples, for two different standard deviations (σ) of the noise, in the third scenario. One can see that it is possible to reveal the secret by applying our approach even against masking.

4. In the fourth scenario, we simulate a leakage that follows the same leakage function as the third scenario, but only for the even values of s. For the odd values of s, we simulate non-relevant points (non-informative points) by random values. This alternation between relevant and dummy samples holds for both shares. Figure 9.13 shows the number of the required traces to reach 80% of success rate according to the number of samples, for two different standard deviations (σ) of the noise, in the last scenario. This figure shows the same trend as Fig. 9.12. The convergence rate with respect to dimensionality (#samples) is similar. But the asymptotic value (#samples to recover the key with 80% of success) is of course higher, especially for high noise.

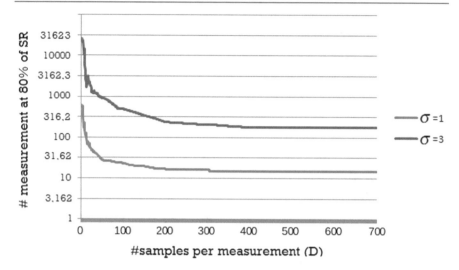

Fig. 9.13 Number of measurements at 80% of success rate (log scale), according to the number of samples, in scenario 4

9.6 Conclusion

In this chapter, we have provided an analytical formula for the template attack in a highly multivariate Gaussian setting. Thanks to the coalescence principle, template attacks can be applied to traces of dimensionality D of several hundreds without prior dimensionality reduction and especially without any effectiveness loss: the success rate increases as D increases. Therefore this study reveals that the high sampling rate of oscilloscopes can help increase the success rate of the attacks.

Furthermore, thanks to a multivariate combination, we extend the approach to the masking-based protected implementations. We also exhibit a spectral approach for template attack which allows an exponential computational improvement in the attack phase (with respect to data bitwidth), which turns out to be a speed up by $32\times$ in the case of AES. Recall that both PoI-based and PCA-based template attacks can straightforwardly benefit from our improvements. Our approach is validated both by simulated and real-world traces.

Random masking is a customary option, but in this case, the countermeasure must be high-order, meaning that each sensitive variable is splitted into multiple (at least two) shares. Attacks therefore become computationally challenging.

In the following chapter, we show that the optimal high-order template attacks (without any multivariate combination) can be expressed under the form of a convolution. This formulation allows for a considerable speed up in their computation thanks to fast Fourier transforms. To further speed up the attack, we also provide an interesting multi-threading implementation of this approach. This strategy naturally applies to template attacks where the leakage of each share is multivariate. We show that this strategy can be adapted to several masking schemes, inherently to the way the splitting is realized.

References

1. Hajra S, Mukhopadhyay D (2015) Reaching the limit of nonprofiling DPA. IEEE Trans CAD Integr Circuits Syst 34(6):915–927
2. Mangard S, Oswald E, Popp T (2006) Power analysis attacks: revealing the secrets of smart cards. Springer. ISBN 0-387-30857-1, http://www.dpabook.org/
3. Chari S, Rao JR, Rohatgi P (2002) Template attacks. In: Kaliski BS, Jr, Koç ÇK, Paar C (eds), Cryptographic hardware and embedded systems - CHES 2002, 4th international workshop, redwood shores, CA, USA, August 13-15, 2002, Revised papers. Lecture notes in computer science, vol 2523. Springer, pp 13–28
4. Choudary O, Kuhn MG (2013) Efficient template attacks. In: Francillon A, Rohatgi P (eds), Smart card research and advanced applications - 12th international conference, CARDIS 2013, Berlin, Germany, November 27-29, 2013. Revised selected papers. LNCS, vol 8419. Springer, pp 253–270
5. Archambeau C, Peeters É, Standaert F-X, Quisquater J-J (2013) Template attacks in principal subspaces. In: CHES. LNCS, vol 4249. Springer, Yokohama, pp 1–14
6. Bär M, Drexler H, Pulkus J. Improved template attacks. In: COSADE, pp 81–89, February 4-5 2010. Darmstadt, Germany. http://cosade2010.cased.de/files/proceedings/cosade2010_paper_14.pdf
7. Elaabid MA, Guilley S (2010) Practical improvements of profiled side-channel attacks on a hardware crypto-accelerator. In: Bernstein DJ, Lange T (eds), Progress in cryptology - AFRICACRYPT 2010, third international conference on cryptology in Africa, Stellenbosch, South Africa, May 3-6, 2010. Proceedings. Lecture notes in computer science, vol 6055. Springer, pp 243–260
8. Fan G, Zhou Y, Zhang H, Feng D (2014) How to choose interesting points for template attacks more effectively? In: Yung M, Zhu L, Yang Y (eds), Trusted systems - 6th international conference, INTRUST 2014, Beijing, China, December 16-17, 2014, Revised selected papers. Lecture notes in computer science, vol 9473. Springer, pp 168–183
9. Zheng Y, Zhou Y, Yu Z, Hu C, Zhang H (2014) How to compare selections of points of interest for side-channel distinguishers in practice? In Lucas Chi Kwong Hui, S. H. Qing, Elaine Shi, and Siu-Ming Yiu, editors, Information and Communications Security - 16th International Conference, ICICS 2014, Hong Kong, China, December 16-17, 2014, Revised Selected Papers, vol 8958 of Lecture Notes in Computer Science, pp 200–214. Springer, 2014
10. Zhang H, Zhou Y (2016) How many interesting points should be used in a template attack? Journal of Systems and Software 120:105–113
11. Debande N, Souissi Y, Elaabid MA, Guilley S, Danger J-L (2012) Wavelet transform based pre-processing for side channel analysis. In 45th annual IEEE/ACM international symposium on microarchitecture, MICRO 2012, workshops procleedings, Vancouver, BC, Canada, December 1–5, 2012. IEEE Computer Society, pp 32–38
12. Maghrebi H, Prouff E (2018) On the use of independent component analysis to denoise side-channel measurements. In: Fan J, Gierlichs B (eds) Constructive side-channel analysis and secure design - 9th international workshop, COSADE 2018, Singapore, April 23-24, 2018, Proceedings. Lecture notes in computer science, vol 10815. Springer, pp 61–81

13. Brier É, Clavier C, Olivier F (2004) Correlation power analysis with a leakage model. In: Joye M, Quisquater J-J (eds), Cryptographic hardware and embedded systems - CHES 2004: 6th international workshop Cambridge, MA, USA, August 11–13, 2004. Proceedings. Lecture notes in computer science, vol 3156. Springer, pp 16–29

14. Lomné V, Prouff E, Roche T (2013) Behind the scene of side channel attacks. In: Sako K, Sarkar P (eds) ASIACRYPT (1). Lecture notes in computer science, vol 8269. Springer, pp 506–525

15. Cochran WG (1934) The distribution of quadratic forms in a normal system, with application to the analysis of covariance. Math Proc Camb Philos Soc 30:178–191

16. Bruneau N, Guilley S, Heuser A, Marion D, Rioul O, Less is more - dimensionality reduction from a theoretical perspective. In Güneysu and Handschuh [99], pp 22–41

17. Coron J-S, Vadnala PK, Giraud C, Prouff E, Renner S, Rivain M (2012) Conversion of security proofs from one model to another: a new issue. In: COSADE. Lecture notes in computer science, May 3–4 2012. Springer, Darmstaft

18. Guilley S, Heuser A, Tang M, Rioul O (2017) Stochastic side-channel leakage analysis via orthonormal decomposition. In: Farshim P, Simion E (eds), Innovative security solutions for information technology and communications - 10th international conference, SecITC 2017, Bucharest, Romania, June 8–9, 2017, Revised selected papers. Lecture notes in computer science, vol 10543. Springer, pp 12–27

19. Bruneau N, Guilley S, Heuser A, Marion D, Rioul O (2017) Optimal side-channel attacks for multivariate leakages and multiple models. J. Cryptograph Eng 7(4):331–341

20. François-Xavier S, Cédric A (2008) Attacks using subspace-based template, to compare and combine power and electromagnetic information leakages. In: CHES. Lecture notes in computer science, vol 5154. Springer, Washington DC, pp. 411–425

21. Jolliffe IT (2002) Principal component analysis. Springer series in statistics. ISBN: 0387954422

22. Zhang H, Zhou Y, Feng D (2015) Mahalanobis distance similarity measure based distinguisher for template attack. Secur Commun Netw 8(5):769–777

23. de Chérisey É, Guilley S, Heuser É, Rioul O (2017) On the optimality and practicability of mutual information analysis in some scenarios. Cryptogr Commun (2017)

24. Prouff E, Rivain M, Bevan R (2009) Statistical analysis of second order differential power analysis. IEEE Trans. Comput 58(6):799–811

25. Messerges TS (2000) Securing the AES finalists against power analysis attacks. Fast Software Encryption'00. Springer, New York, pp 150–164

26. Williams VV (2012) Multiplying matrices faster than coppersmith-winograd. In: STOC'12 Proceedings of the forty-fourth annual ACM symposium on theory of computing, New York, USA — 19 - 22, 2012, pp 887–898

27. Jean-Sébastien C, Louis G (2000) Boolean on, masking arithmetic, against differential power analysis. In: CHES. Lecture notes in computer science, vol 1965. Springer, Worcester, pp 231–237

28. Messerges TS, Dabbish EA, Sloan RH (2002) Examining smart-card security under the threat of power analysis attacks. IEEE Trans. Comput 51(5):541–552

29. Lerman L, Poussier R, Bontempi G, Markowitch O, Standaert F-X. Template attacks vs. machine learning revisited (and the curse of dimensionality in side-channel analysis). In: Mangard and Poschmann [136], pp 20–33

30. Messerges TS (2000) Securing the AES finalists against power analysis attacks. In: B. Schneier (ed), Fast software encryption, 7th international workshop, FSE 2000, New York, NY, USA, April 10-12, 2000, proceedings. Lecture notes in computer science, vol 1978. Springer, pp 150–164

31. Clavier C, Coron J-S, Dabbous N (2000) Differential power analysis in the presence of hardware countermeasures. In: Koç ÇK, Paar C (eds), CHES. Lecture notes in computer science, vol 1965. Springer, pp 252–263

Spectral Approach to Process the High-Order Template Attack Against any Masking Scheme

10

10.1 Introduction

Cryptographic devices manage secret keys, which must be protected against extraction. One stealthy attack consists in the analysis of side-channel leakage. As a countermeasure, cryptographic computations can be randomly masked.

The extent of protection bestowed by random masking has been the topic of extensive research. Profiling attacks have proven to be the most efficient, as they model the leakage with great accuracy and subsequently resort to maximizing the key extraction likelihood. Such attacks are known as *template attacks* [2]. When confronted to randomly masked implementations, they can be tailored, as explained, for instance, in [1]. This paper shows that each share (random split of sensitive variables) can have its leakage profiled, and the resulting high-order template attack recombines all the profiles in order to maximize the key extraction probability.

10.1.1 Related Works

The seminal paper of Template Attacks (TA) is based on the maximum likelihood principle [2]. Since the importance of masking, several extensions of the TA have been introduced to defeat (or to assess) this countermeasure [1,3].

In 2014, the Higher Order Optimal Distinguishers (HOOD) has been introduced against the masking countermeasure, whatever its order [4]. It is dubbed optimal in that it consists of computing a maximum likelihood. Nevertheless, its complexity is exponentially linked to the masking order. Thereby, their authors considered only a monovariate distribution (one-time sample per share). In practice, side-channel traces are multivariate, because oscilloscopes capture waveforms for each sensitive variable manipulation. Therefore, the HOOD as per [4] is far from being the most suitable distinguisher (for one-time sample per share), since we expect that an increase of the number of time samples per share can significantly improve the success rate of the attack [5].

© The Author(s), under exclusive license to Springer Nature Switzerland AG 2021
M. Ouladj and S. Guilley, *Side-Channel Analysis of Embedded Systems*,
https://doi.org/10.1007/978-3-030-77222-2_10

Recently, the optimal side-channel attacks for multivariate leakages and multiple models are introduced [6]. Its inconveniences are twofold. First, this modeling is over-constrained, as there is an additional assumption of a relation between the templates (a coefficient α between each pair of templates' mean). Second, its computation has an exponential complexity according to the masking order.

Subsequently, a Taylor expansion of the maximum likelihood distinguisher has been introduced to reduce its complexity [7]. In this paper, the processing is simplified in pre-computations (rounded at a given order, at the expense of the template precision). The online attack computations are therefore efficient.

More recently, a new approach is introduced in [8, Sect. 4.4.2]. Its core idea is to combine the leaking shares into one *artificial* trace (inspired from the *normalized* product combination [9] of each share's leakage), for carrying out a template attack efficiently in terms of implementation complexity. Its inconvenient is the fact that this combination could lose some information about the leakage, especially for low-noise devices [10].

The work presented in our chapter shows that neither the rounding of attacks nor the shares combination is necessary for computational reasons. Besides, this result holds for any masking scheme.

10.1.2 Contributions

High-order masking schemes consist of randomly splitting each sensitive variable into several shares, so as to make attacks more difficult (it can be proven that attacks are exponentially complex with the number of shares, in terms of number of traces [11,12]). Those shares can be linearly combined to carry out computations on them (while remaining protected by random masking). Typically, this property also allows to demask at the end of the computation, when all shares need to be aggregated to recover a ciphertext.

The same property is used by an attacker to mount so-called high-order attacks; namely, the attacker combines shares so that jointly they do contain information on the sensitive variables. The natural combination consists in undoing the sharing, i.e., to conduct a linear combination of the individual shares leakage. For masking of order d, each tuple of less than or equal to d shares consists of independent variables (i.e., a free family). On the contrary, a tuple of $N = d + 1$ shares has exactly one linear relationship.

In this chapter, we leverage this basic noting to conceptualize high-order attacks as a convolution product over the d-dimensional mask space. This convolution stands out in a straightforward manner for Boolean masking. But **it can also be adapted to other masking styles, by the introduction of the relevant variable change**. The linear operation used in the masking scheme impacts the type of convolution, which we capture using the corresponding group law.

This new vision of masking schemes can therefore take advantage of the considerable wealth of *character theory*. In particular, the computation of the side-channel distinguishers can be optimized thanks to Plancherel identity (arising from properties of Fourier transforms).

The contributions of this chapter are therefore to:

- Quantify the speed up in distinguisher computation;
- Show that this speed up is particularly adapted to the optimal distinguisher (therefore: two advantages for the attacker—minimizing the required number of traces to extract the key [*optimal distinguisher using maximum likelihood*] and minimizing the attack elapsed time and required memory space [*optimal distinguisher computational complexity*]);
- Provide an interesting multi-threading implementation of our approach;
- Apply the attack in the context of multivariate sub-traces for the leakage of each share;
- Illustrate quantitatively this new attack paradigm on (very) high-order masking schemes, including **six** styles (Boolean, IPM, DSM, Polynomial DSM, RSM, leakage squeezing);
- Introduce a new multi-share extension of DSM (along with the optimal attack on it).

10.1.3 Outline

The rest of the chapter is structured as follows. Mathematical notions and results useful to present the chapter contributions are gathered in Sect. 10.2. The algorithmic optimization of the high-order multivariate distinguisher is presented in Sect. 10.3, on the simple example of Boolean Masking. Its adaptation per masking type is the topic of Sect. 10.4. In particular, Boolean, i.e., additive, multiplicative, and affine masking schemes, are addressed in Sect. 10.4.1 (which covers all these schemes altogether exploiting the fact they are instances of the IPM scheme which generalizes them). DSM and its multi-share extension are treated, respectively, in Sects. 10.4.2 and 10.4.3. The polynomial extension of DSM, which (as DSM) allows for both fault detection and side-channel attack protection, is the topic of Sect. 10.4.4. The Rotating Substitution-box Masking is analyzed in Sect. 10.4.5. Eventually, leakage squeezing is addressed in Sect. 10.4.6. Validation of the theory on real traces is the topic to Sect. 10.5. Eventually, conclusions and perspectives are given in Sect. 10.6. The Appendix 10.7 explains the flaw of the straightforward extension of DSM to the multi-share case, and also provides the subsequent fix.

10.2 Preliminaries

10.2.1 Linear Algebra and Linear Codes

Definition 10.1 (*Complementary subspaces*) Let n be a non-zero integer. Let $\mathbb{C}$ and $\overline{\mathbb{C}}$ be two subspaces (seen as linear codes) of a vector space $\mathbb{F}_2^n$. By definition $\mathbb{C}$ and $\overline{\mathbb{C}}$ are complementary subspaces, if and only if, any element of the vector space $\mathbb{F}_2^n$

can be decomposed in a unique way as a sum of two elements of $\mathbb{C}$ and $\overline{\mathbb{C}}$. We write $\mathbb{F}_2^n = \mathbb{C} \oplus \overline{\mathbb{C}}$.

Remark 10.1 The supplementary subspace of a given subspace is not unique and it is not an orthogonal subspace in the general case. If a linear code $\mathbb{C}$ has an orthogonal supplementary code (a dual code denoted $\mathbb{C}^\perp$), then $\mathbb{C}$ is called a Linear Complementary Dual (LCD) code, and given their dimension, we have that the so-called hull $\mathbb{C} \cap \mathbb{C}^\perp$ of $\mathbb{C}$ is trivial, namely $\mathbb{C} \cap \mathbb{C}^\perp = \{0\}$ [13,14].

By the rank-nullity theorem, if the dimension of $\mathbb{C}$ equals $n_0 < n$, then the dimension of any supplementary subspace $\overline{\mathbb{C}}$ is equal $n - n_0$.

Let us consider generating matrices G and $\overline{G}$ of, respectively, $\mathbb{C}$ and $\overline{\mathbb{C}}$. We denote $\mathbb{C} = \text{span}(G)$ and $\overline{\mathbb{C}} = \text{span}(\overline{G})$. Then every vector $z \in \mathbb{F}_2^n$ can be written in a unique way as $z = z_1 G \oplus z_2 \overline{G}$, where $z_1 \in \mathbb{F}_2^{n_0}$ and $z_2 \in \mathbb{F}_2^{n-n_0}$.

Definition 10.2 (*Minimum distance*) The minimum distance $d_\mathbb{C}$ of a linear code $\mathbb{C}$ is the smallest Hamming weight of its non-zero codeword.

Definition 10.3 (*Dual distance*) The dual distance $d_{\mathcal{D}}^\perp$ of a linear code $\mathcal{D}$ is $d_{\mathcal{D}}^\perp = d_{\mathcal{D}^\perp}$.

10.3 Higher Order Template Attack

This section illustrates the approach (multivariate leakage and computational speed-up) on Boolean Masking. Generalization to other masking schemes is the topic of Sect. 10.4.

10.3.1 High-Order Boolean Masking

In this section, we introduce the notations. Let us assume an attacker knows the plaintext t and guesses a constant secret key k. They are inputted in a cryptographic algorithm, and their combination gives rise to a *sensitive variable* z. For example, $z(t, k) = \text{Sbox}(t \oplus k)$, in the case of the attack of a substitution box (in block ciphers such as AES or PRESENT). In general, the function $k \mapsto z(t, k)$ is an injection, for all t.

The sensitive variable z is the target of side-channel attacks. In masking schemes, each z is randomly splitted. Namely, dth higher order masking consists in computing each sensitive variable $z(t, k)$ separately in $N = d + 1$ shares.

For example, in dth order Boolean masking scheme, z is splitted in $(z^{(0)}, \ldots, z^{(d)})$, such that

$$z = \bigoplus_{w=1}^{d} z^{(w)}. \tag{10.1}$$

In Eq. (10.1), the $d + 1$ shares are traditionally randomly chosen from d random numbers called masks m_w, with $1 \leq w \leq d$, according to

- $z^{(1)} = m_1$,
- $z^{(2)} = m_2$,
- $\ldots$
- $z^{(d)} = m_d$,
- $z^{(0)} = z \oplus \bigoplus_{w=1}^{d} z^{(w)}$—recall Eq. (10.1).

Besides, in Eq. (10.1), the operation is the bitwise XOR, denoted as "$\oplus$".

In general, such masking adapts to other situations. Typically, one just needs the combination to be a group operation. In the rest of this section, we shall consider that this group is additive. We denote it as $(\mathbb{S}, +)$. Notice that multiplicative groups also allow to derive multiplicative masking; Therefore, the "$+$" sign shall be understood broadly, as the group $\mathbb{S}$ inner operation. Examples of groups $\mathbb{S}$ are:

1. $(\mathbb{F}_2^n, \oplus)$, used in Boolean masking, or
2. $(\mathbb{F}_{2^n}, +)$, such that $+$ is the 2^n-modular addition, which is used in arithmetic masking.

Some cryptographic algorithms make use both Boolean and arithmetic maskings, thus the two masking schemes shall cohabit, and masking conversions shall be implemented [15].

10.3.2 Attack on High-Order Boolean Masking

Template attacks have been introduced as multivariate attacks on unprotected devices in [2]. Their extension to masked implementation has been suggested in [1,3]. However, template attacks on masking schemes where each share is profiled independently are not clearly described in the state of the art. We discuss it in this chapter, which gives the most general attack compared to the existing literature.

During the profiling phase, the adversary can see the leakage measurement X_q as $(d + 1)$ disjoint sub-traces denoted by $X_q^{(w)}, 0 \leq w \leq d$. Each sub-trace $X_q^{(w)}$ is a sub-trace of length $D^{(w)}$ and corresponds to the leakage of share $z^{(w)}$. We assume that for each w, $0 \leq w \leq d$, the adversary estimates the multivariate Probability Density Function (PDF) of $X_q^{(w)}$, such that she profiles the device consumption of $D^{(w)}$ samples for the wth share.

Thanks to the profiling result, the adversary can estimate the probability $p(X_q^{(w)}|z^{(w)})$, for each manipulated data $z^{(w)}$. We recall that $z^{(w)}$, for $1 \leq w \leq d$, is equal to mask m_w, and $z^{(0)}$ is equal to $z \oplus m_1 \oplus \cdots \oplus m_d$.

According to [4, Theorem 7], the $(d + 1)$th-order optimal distinguisher is (for simplicity, we assume that all the mask values have the same probability):

$$\mathscr{D}_{opt}^d = \underset{k}{\operatorname{argmax}} \prod_{q=1}^{Q} \sum_{m_1,\ldots,m_d \in \mathbb{S}} p(X_q^{(1)}|m_1) \ldots p(X_q^{(d)}|m_d) p(X_q^{(0)}|z(t_q,k) \oplus \bigoplus_{w=1}^{d} m_w). \quad (10.2)$$

This equation is the maximum likelihood estimator. It takes as inputs:

- the known plaintexts $(t_q)_{1 \leq q \leq Q}$ and
- the leakage sub-traces $(X_q^{(0)}, \ldots, X_q^{(d)})_{1 \leq q \leq Q}$,

and returns the most likely key $\hat{k}$ according to the relationship between z, t, and k.

In fact, in [4], the HOOD is studied by tacking only one sample per share. In our study, we generalize it by assuming $D^{(w)}$ samples for the wth sub-trace, $0 \leq w \leq d$, and considering this sub-trace as a random vector. That means, we generalize the HOOD in same way when one goes from the first-order DPA (one sample per trace to compute the distinguisher) to the (first-order) template attack (D samples per trace but concerning only one sensitive variable z). Here, we migrate from the higher order DPA to the higher order template attack.

The attack setup is depicted in Fig. 10.1. In this figure, the leakage of each share is represented as a bivariate Gaussian, for the sole sake of illustration. However, this setup works well for higher multiplicity, and even for different multiplicities $D^{(w)} > 0$ for each share w, $0 \leq w \leq d$. This figure represents the attack on Boolean Masking with $N = d + 1$ shares. Adaptation to other masking schemes is the topic of Sect. 10.4.

To compute our distinguisher efficiently, one can see that

$$\mathscr{D}_{opt}^d = \underset{k}{\operatorname{argmax}} \prod_{q=1}^{Q} \operatorname{Sum}_q^{\mathsf{BM}}(z(t_q,k)) = \underset{k}{\operatorname{argmax}} \sum_{q=1}^{Q} \log \operatorname{Sum}_q^{\mathsf{BM}}(z(t_q,k)),$$

where $\operatorname{Sum}_q^{\mathsf{BM}}$ is a pseudo-Boolean function (meaning a real function defined over Boolean vectors) of 2^n values, equals to

$$\operatorname{Sum}_q^{\mathsf{BM}}(z) = \sum_{m^{(1)},\ldots,m^{(d)} \in \mathbb{S}} p(X_q^{(1)}|m^{(1)}) \ldots p(X_q^{(d)}|m^{(d)}) p(X_q^{(0)}|z \oplus m^{(1)} \oplus \cdots \oplus m^{(d)}) \quad (10.3)$$

$$= \sum_{m^{(1)} \in \mathbb{S}} \cdots \sum_{m^{(d)} \in \mathbb{S}} p(X_q^{(1)}|m^{(1)}) \ldots p(X_q^{(d)}|m^{(d)}) p(X_q^{(0)}|z \oplus m^{(1)} \oplus \cdots \oplus m^{(d)})$$

$$= p(X_q^{(1)}|.) \otimes \cdots \otimes p(X_q^{(d)}|.) \otimes p(X_q^{(0)}|.)(z), \quad (10.4)$$

such that $\otimes$ is the convolution product relatively to the law $\oplus$ in the group $(\mathbb{S}, \oplus)$. Clearly, the property that the value of z can be recovered as $z = z^{(0)} - \sum_{w=1}^{d} z^{(w)}$ is central to apparition of the convolution. This property is specific to the Boolean

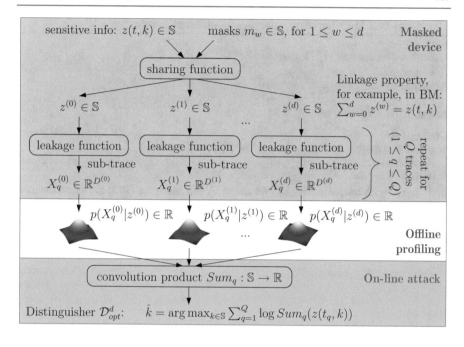

Fig. 10.1 Synoptic of the multivariate high-order attack, applied on dth-order Boolean Masking (BM)

Masking, but extends to all masking schemes, as will be shown in Sect. 10.4. Indeed, in all masking schemes, there is some sort of relationship between the shares $z^{(0)}, \ldots, z^{(d)}$ which allows to recover sensitive value $z(t, k)$.

The convolution which we exhibit in Sum_q^{BM}, namely in Eq. (10.4), allows us to compute, for every q, all the 2^n values of $Sum_q^{\mathsf{BM}}(z), z \in \mathbb{F}_2^n$, at once. Indeed, the Fast Walsh-Hadamard relies on a butterfly algorithm of complexity $nd2^n$, whereas the naïve computation has complexity $2nd$ (direct multiple summation over $\mathbb{F}_2^n \times \ldots \times \mathbb{F}_2^n$, as per Eq. (10.3)).

The overall attack is described in Algorithm 10.1). It can be noticed that line 9 assumes that the function $k \mapsto z(t, k)$ is injective.

Thanks to the formalization in a convolution, one can compute the Sum_q^{BM} of all the possible z at once. So, the overall computation complexity of the H-O Template Attack distinguisher (Eq.(10.2)) is $Qdn2^n$ instead of $Q2nd$. Note that our computed distinguisher is a generalization of HOOD, in which only $(D^{(w)} = 1)_{w=0,\ldots,d}$ is considered. But our processing is $\frac{2^{nd}}{dn2^n}$ times faster.

Also, let us precise that the optimization of HOOD has been proposed in [7], in which the computationally intensive sums over all masks are simplified in precomputations (rounded at a given order). The online attack computations are therefore efficient. But the work presented in this chapter shows that thanks to Fourier transform, the rounding of attacks is not necessary (there are no longer computational

---

Algorithm 10.1 H-O Template Attacks with efficient processing.

Input   :
- Templates: for example $(Y_z^{(w)})_{0 \leq z < 2^n}$ and $\Sigma^{(w)}$;
- Sub-traces: $(X_q^{(0)}, \ldots, X_q^{(d)})_{1 \leq q \leq Q}$;
- Plaintexts: $T = (t_q)_{1 \leq q \leq Q}$,

Output   : The most likely key $\mathscr{D}_{opt}^d \in \mathbb{F}_2^n$.

1 $Distinguisher \leftarrow \{0, \ldots 0\}$;   // A vector of 2^n zeros
2 **for** $q \leftarrow 1$ **to** Q **do**
3 $\quad$ $Product \leftarrow \{1, \ldots 1\}$;   // A vector of 2^n ones
4 $\quad$ **for** $w \leftarrow 0$ **to** d **do**
5 $\quad\quad$ $p(X_q^{(w)}|z)$ is computed from $X_q^{(w)}$ for all $z \in \mathbb{F}_2^n$ (e.g., using Eq. (10.5) and template means
   $\quad\quad$ $Y_z^{(w)}$ and their common covariance matrix $\Sigma^{(w)}$) $FFT(p(X_q^{(w)}|.))$;   // In-place FFT on
   $\quad\quad$ 2^n-valued vector $p(X_q^{(w)}|.) = \{p(X_q^{(w)}|z), z \in \mathbb{F}_2^n\}$
   $\quad\quad$;   // The lines 5 and 5 could be processed in parallel as will be shown in Sec. 10.5.3
6 $\quad\quad$ $Product(.) \leftarrow Product(.) \bullet p(X_q^{(w)}|.)$
7 $\quad$ $Sum_q \leftarrow FFT^{-1}(Product(.))$;   // Computation of convolution
8 $\quad$ **for** $k \in \mathbb{F}_2^n$ **do**   // Pigeonholing spectral components based on the plaintext $t_q \in \mathbb{F}_2^n$ value
9 $\quad\quad$ $Distinguisher(k) \leftarrow Distinguisher(k) + \log Sum_q(z(t_q, k))$;   // Permutation of Sum_q
   $\quad\quad$ indices

10 **return** $\underset{k \in \mathbb{F}_2^n}{\mathrm{argmax}}\ Distinguisher(k)$

---

reasons). Hence, one major contribution is to show how to compute efficiently and exactly in high-order (possibly multivariate) contexts.

10.3.3   Computing the Template Profile Functions $p(X_q^{(w)}|.)$

First, we notice that the *coalescence* principle [8] cannot be taken advantage of because $\prod_{q=1}^{Q} \sum_{m_1, \ldots, m_d \in \mathbb{S}}$ is different from $\sum_{m_1, \ldots, m_d \in \mathbb{S}} \prod_{q=1}^{Q}$. Intuitively, this means that averaging traces would result in the masking countermeasure canceling the information.

For each possible value z of the wth share, the adversary should compute $p(X_q^{(w)}|z)$. Thanks to the profiling phase, the adversary estimates this probability. For example, let us assume that the wth random vector follows a Gaussian distribution PDF ($X^{(w)} \sim \mathcal{N}(Y^{(w)}, \Sigma^{(w)})$, as suggested in the seminal TA paper [2]), where $Y^{(w)}$ is the mean of the signal and $\Sigma^{(w)}$ the noise covariance matrix. In this case, one has

$$p(X_q^{(w)}|z) = \frac{1}{\sqrt{(2\pi)^{D^{(w)}} |\det \Sigma^{(w)}|}} \exp\left(-\frac{1}{2}\left(X_q^{(w)} - Y_z^{(w)}\right)^{\mathsf{T}} \Sigma^{(w)-1}\left(X_q^{(w)} - Y_z^{(w)}\right)\right), \quad (10.5)$$

where T denotes the matrix transposition operator. For the sake of clarity, we recall the dimensions:

- $X_q^{(w)}$ and $Y_z^{(w)}$ are $D^{(w)} \times 1$,
- $\Sigma^{(w)}$ is $D^{(w)} \times D^{(w)}$.

One can remove the factor independent from the key (i.e., from z), and keep only

$$
\begin{aligned}
p'(X_q^{(w)}|z) &= exp\left((X_q^{(w)})^\mathsf{T} \Sigma^{(w)-1} Y_z^{(w)} - \frac{1}{2}(Y_z^{(w)})^\mathsf{T} \Sigma^{(w)-1} Y_z^{(w)} \right) \\
&= exp\left(((X_q^{(w)})^\mathsf{T} - \frac{1}{2}(Y_z^{(w)})^\mathsf{T}) \Sigma^{(w)-1} Y_z^{(w)} \right) \\
&= exp\left((X_q^{(w)} - \frac{1}{2}Y_z^{(w)})^\mathsf{T} \Sigma^{(w)-1} Y_z^{(w)} \right).
\end{aligned}
$$

It is noteworthy that the vectors $\Sigma^{(w)-1} Y_z^{(w)}$ can be pre-computed **only once** at the end of the profiling phase.

10.3.4   Equivalent Multivariate Signal-to-Noise Ratio (SNR)

The presentation of multivariate probability distribution at each share (recall (10.5)) shows that the leakage can take advantage of dimensionality reduction. The transformation is that already described in the paper [16, Theorem 8]. This equivalent SNR can thus be deduced exactly with the formula given in [16, Corollary 4], namely:

$$
Y_z^{(w)\mathsf{T}} \Sigma^{-1} Y_z^{(w)}, \text{ for all } w.
$$

10.4   Type of Fourier Transform per Masking Scheme

In this section, we detail the customizations which shall be performed to adapt the principle of fast multivariate high-order template attack (Sect. 5.1) to several masking types.

For the Boolean masking, the group law is simply the XOR operation between the $\mathbb{F}_2^n$ elements. Thereby, the group is $(\mathbb{F}_2^n, \oplus)$. Subsequently, the corresponding Fourier transform is simply the Walsh-Hadamard Transform (WHT). Similarly, for a certain arithmetic masking scheme [17], the group law is the modular addition operation between the $\mathbb{F}_{2^n}$ elements. Subsequently, the corresponding Fourier transform is the Cyclical Fourier Transform.

To generalize our approach for other masking schemes, we study the form taken by the term $Sum_q(z)$ defined in Eq. (10.3), and how it shall be adapted accordingly. This expression will be specialized with the abbreviation of the masking scheme, for instance, $Sum_q^{\mathsf{IPM}}(z)$ for the IPM masking scheme.

10.4.1 Type of Fourier Transform for Inner Product Masking (IPM) Scheme

Let us first focus on the Inner Product Masking (IPM) [18–20]. This choice is due to the fact that the IPM is a generalisation of several simple masking schemes (i.e., Boolean Sect. 4.2.1, additive, multiplicative Sect. 4.2.2, and affine Sect. 4.2.3 masking schemes) Sect. 4.2.8.

Instead of simply splitting every sensitive value as the sum (i.e., $\oplus$) of random shares, IPM consists in decomposing every secret in an inner product between $d + 1$ random values and a $d+1$-dimensional constant public vector [21]. Formally, the designers choose a constant public vector $L = (l_0, \ldots, l_d) \in \mathbb{F}_{2^n}^{d+1} \backslash \{0\}$. Therefore, all field element l_w is invertible. In practice, and without loss of generality, l_0 is chosen equals 1, for performance reasons. Masking a sensitive value z by IPM means computing $d + 1$ shares all in $\mathbb{F}_{2^n}$ and denoted $z^{(0)}, \ldots z^{(d)}$, such that $z = \sum_{w=0}^{d} l_w z^{(w)}$.

Let us consider as in practice $l_0 = 1$ (this condition is not restrictive). Thereby, one has $z^{(0)} = z - \sum_{w=1}^{d} l_w z^{(w)}$ So, the adversary needs to combine the $d+1$ leaking shares to lead a successful H-O attack. According to above the optimal adversary has to maximize,

$$\mathcal{D}_{opt}^d = \underset{k}{\operatorname{argmax}} \prod_{q=1}^{Q} \sum_{z^{(1)}, \ldots, z^{(d)} \in \mathbb{F}_{2^n}} p(X_q^{(1)}|z^{(1)}) \ldots p(X_q^{(d)}|z^{(d)}) p(X_q^{(0)}|z - \sum_{w=1}^{d} l_w z^{(w)}). \quad (10.6)$$

But,

$$
\begin{aligned}
Sum_q^{\mathsf{IPM}}(z) &= \sum_{(z^{(1)}, \ldots, z^{(d)}) \in \prod_{w=1}^{d} \mathbb{F}_{2^n}} p(X_q^{(1)}|z^{(1)}) \ldots p(X_q^{(d)}|z^{(d)}) p(X_q^{(0)}|z - \sum_{w=1}^{d} l_w z^{(w)}) \\
&= \sum_{(z^{(1)}, \ldots, z^{(d)}) \in \prod_{w=1}^{d} \mathbb{F}_{2^n}} p(X_q^{(1)}|l_1^{-1} l_1 z^{(1)}) \ldots p(X_q^{(d)}|l_d^{-1} l_d z^{(d)}) p(X_q^{(0)}|z - \sum_{w=1}^{d} l_w z^{(w)}) \\
&= \sum_{(z^{(1)}, \ldots, z^{(d)}) \in \prod_{w=1}^{d} l_w \mathbb{F}_{2^n}} p(X_q^{(1)}|l_1^{-1} z^{(1)}) \ldots p(X_q^{(d)}|l_d^{-1} z^{(d)}) p(X_q^{(0)}|z - \sum_{w=1}^{d} z^{(w)}) \\
&= \sum_{(z^{(1)}, \ldots, z^{(d)}) \in \prod_{w=1}^{d} l_w \mathbb{F}_{2^n}} p'(X_q^{(1)}|z^{(1)}) \ldots p'(X_q^{(d)}|z^{(d)}) p(X_q^{(0)}|z - \sum_{w=1}^{d} z^{(w)}),
\end{aligned}
$$

where $p'(X_q^{(w)}|z^{(w)}) = p(X_q^{(w)}|l_w^{-1} z^{(w)})$, for all $w = 1 \ldots d$. In fact, for any $l_w \neq 0$, one has $l_w \mathbb{F}_{2^n} = \mathbb{F}_{2^n}$ because $x \in \mathbb{F}_{2^n} \mapsto l_w x$ is a bijection.

Subsequently, the Sum_q^{IPM} becomes a convolution relatively to the addition law in group $\mathbb{F}_{2^n}$. Thereby, it can be processed efficiently thanks to the (Cyclical) Fourier Transform (CFT),

$$Sum_q^{\mathsf{IPM}}(z) = \left(p'(X_q^{(1)}|.) \otimes \cdots \otimes p'(X_q^{(d)}|.) \otimes p(X_q^{(0)}|.) \right)(z),$$

$$= CFT^{-1} \left(CFT(p'(X_q^{(1)}|.)) \bullet \cdots \bullet CFT(p'(X_q^{(d)}|.)) \bullet CFT(p(X_q^{(0)}|.)) \right)(z).$$

So, the overall computation complexity of the H-O Template Attack distinguisher against the IPM is as expected above (Eq.(10.2)) $Qdn2^n$ instead of $Q2nd$.

10.4.2   Type of Fourier Transform for Direct Sum Masking (DSM) Scheme

Since the IPM is a particular case of DSM [22], let us further generalize our approach for DSM Sect. 4.2.9. DSM is an error correction code-based masking scheme introduced in [23,24]. Unlike IPM (which is defined over the field $\mathbb{F}_{2^n}$), DSM is defined over the spacevector $\mathbb{F}_2^n$.

To formally describe DSM, let $\mathbb{C}$ (resp. $\mathscr{D}$) be a linear code of dimension n (resp. n') and G (resp. H) be its generating matrix. The codes $\mathbb{C}$ and $\mathscr{D}$ are chosen to be complementary. This means that square matrix $\begin{pmatrix} G \\ H \end{pmatrix}$ has full rank. and we denote J (resp. K) the $(n + n') \times n$ (resp. $(n + n') \times n'$) matrix such that

$$\begin{pmatrix} G \\ H \end{pmatrix}^{-1} = (J \ K). \tag{10.7}$$

When $\mathbb{C}$ and $\mathscr{D}$ are also dual, in addition to be complementary, then such codes have been studied by Massey [13]. In this case, $J = G^T(GG^T)^{-1}$ and $K = H^T(HH^T)^{-1}$. But in general, $\mathbb{C}$ and $\mathscr{D}$ are not required to be orthogonal. Using these two complementary codes, DSM consists in

- first encoding the sensitive value $z \in \mathbb{F}_2^n$ in the n-dimension code $\mathbb{C}$ by processing $zG \in \mathbb{F}_2^{n+n'}$,
- second encoding a random value $m \in \mathbb{F}_2^{n'}$ in the n'-dimension code $\mathscr{D}$ by processing $z^{(1)} = mH \in \mathbb{F}_2^{n+n'}$,
- third and finally, masking z by processing $z^{(0)} = zG \oplus mH$.

The advantage of this scheme from the designer's point of view is the efficient method of de-masking (namely, a projection) since $\mathbb{C}$ and $\mathscr{D}$ are complementary: the sensitive variable is recovered by computing $z = z^{(0)}J$.

Even if this masking scheme has $d_{\mathscr{D}}^{\perp} - 1$ (i.e., $d_{\mathbb{C}} - 1$ when $\mathscr{D} = \mathbb{C}^{\perp}$) as a masking order (at bit level), the corresponding Sum_q is simply

$$\forall z \in \mathbb{C}, \quad Sum_q^{\mathsf{DSM}}(z) = \sum_{m \in \mathbb{F}_{2^{n'}}} p(X_q^{(1)}|m)p(X_q^{(0)}|zG \oplus mH). \tag{10.8}$$

Notice that

$$Sum_q^{\text{DSM}}(z) = \sum_{z^{(1)} \in \mathbb{F}_2^{n+n'}} 1_{\mathscr{D}}(z^{(1)}) p'(X_q^{(1)}|z^{(1)}) p(X_q^{(0)}|zG \oplus z^{(1)}), \tag{10.9}$$

where $p'(X_q^{(1)}|mH) = p(X_q^{(1)}|m)$ The computation according to Eq. (10.9) is not efficient because its complexity is $\mathscr{O}((n + n')2^{n+n'})$. But, the straightforward processing of the Sum_q^{DSM} (according to Eq. (10.8)) is not so difficult. It has an overall complexity of $\mathscr{O}(2^{n+n'})$. In fact, one cannot computer faster than Eq. (10.8) because zG and mH are not belonging to the same subspace.

10.4.3 Multi-share DSM (MS-DSM)

In this section, we aim at improving the security of DSM. For more details, reader is invited to Sect. 4.2.9.1 and Appendix 10.7.

Rationale for the Straightforward Extension of DSM
One option consists in increasing the dual distance of the code $\mathscr{D}$, which comes at the expense of longer codewords (recall that in DSM, there is only one share, namely $z^{(0)}$). Indeed, at bit level,

Lemma 10.1 (Proposition 1 of [22]). *The resistance order of DSM is equal to the dual distance of code $\mathscr{D}$ minus the number one, that is $d_{\mathscr{D}}^{\perp} - 1$.*

At some point, these codewords will have a length which is incompatible with the host processor register size, Or, with the size of the memories, they address.
   Therefore, a second option shall be envisioned. It consists of

- starting from a regular DSM protection,
- but by adding further masks within $\mathscr{D}$.

Say d masks are used. They consist in the encoding of d random values $m_1, \ldots, m_d$ in the same n'-dimension code $\mathscr{D}$ by processing $(z^{(w)} = m_w H)_{w=1,\ldots,d}$. Therefore, the splitting is as follows:

$$(z, m_1, \ldots, m_d) \in \mathbb{F}_2^n \times (\mathbb{F}_2^{n'})^d \mapsto (z^{(0)}, z^{(1)}, \ldots, z^{(d)}) = (zG \oplus \bigoplus_{w=1}^d m_w H, m_1, \ldots, m_d).$$
$$\tag{10.10}$$

The advantage of DSM in terms of ease of demasking is preserved: as all masks belong to $\mathscr{D}$, so is their sum, hence a single projection on $\mathbb{C}$ allows to remove them all at once. In practice, this means that $z = z^{(0)} J$. Furthermore, the different shares can now be processed independently, which makes mapping on the processor registers easy. Specifically, using DSM, the register bitwidth shall be enough to accommodate $n + n'$ bits, but in MS-DSM, one can see in Eq. (10.10) that the masked value is broken in $d + 1$ different values.

From an adversary point of view, our approach is useful for carrying out the maximum likelihood-based (optimal) distinguisher. Against such scheme, the $Sum_q^{\text{MS-DSM}}$ becomes

$$Sum_q^{\text{MS-DSM}}(z)$$

$$= \sum_{(m_1,\ldots,m_d) \in \Pi_{w=1}^d \mathbb{F}_2^{n'}} p(X_q^{(1)}|m_1) \ldots p(X_q^{(d)}|m_d) p(X_q^{(0)}|zG \oplus m_1 H \cdots \oplus m_d H)$$

$$= \sum_{(m_1,\ldots,m_d) \in \Pi_{w=1}^d \mathbb{F}_2^{n'}} p(X_q^{(1)}|m_1) \ldots p(X_q^{(d)}|m_d) p_z(X_q^{(0)}|(0 \oplus m_1 \cdots \oplus m_d)H)$$

$$= \sum_{(m_1,\ldots,m_d) \in \Pi_{w=1}^d \mathbb{F}_2^{n'}} p(X_q^{(1)}|m_1) \ldots p(X_q^{(d)}|m_d) p_z'(X_q^{(0)}|0 \oplus m_1 \cdots \oplus m_d)$$

$$= p(X_q^{(1)}|.) \otimes \cdots \otimes p(X_q^{(d)}|.) \otimes p_z'(X_q^{(0)}|.)(0),$$

where $p_z'(X_q^{(0)}|m) = p_z(X_q^{(0)}|mH) = p(X_q^{(0)}|zG \oplus mH)$, for $m \in \mathbb{F}_2^{n'}$.

Subsequently, the sum $Sum_q^{\text{MS-DSM}}$ becomes a convolution relatively to the $\oplus$ law in the group $(\mathbb{F}_2^{n'}, \oplus)$. Thereby, it can be processed efficiently for each $zG \in \mathcal{C}$, thanks to the WHT, as shown below:

$$Sum_q^{\text{MS-DSM}}(z)$$

$$= WHT^{-1} \left(WHT(p(X_q^{(1)}|.)) \bullet \cdots \bullet WHT(p(X_q^{(d)}|.)) \bullet WHT(p_z'(X_q^{(0)}|.)) \right)(0)$$

$$= \frac{1}{2^{\frac{n'}{2}}} \sum_{m \in \mathbb{F}_2^{n'}} \left(WHT(p(X_q^{(1)}|.)) \bullet \cdots \bullet WHT(p(X_q^{(d)}|.)) \bullet WHT(p_z'(X_q^{(0)}|.)) \right)(m).$$

One can process the product function $WHT(p(X_q^{(1)}|.)) \bullet \cdots \bullet WHT(p(X_q^{(d)}|.))$ only once (for each q) since it is independent from z. Also, one can ignore the factor $\frac{1}{2^{n'/2}}$ since it is independent from k. Its overall complexity is about $Qdn2^n$.

So, the overall computation complexity of the H-O Template Attack distinguisher against the so-called MS-DSM is slightly different from that expected above (Eq. (10.2)). . It is about $Qdn'2^{n'} + Qn'2^{n+n'}$ instead of $Qdn2^n$. It is noticeable that this complexity is still better than that of the naïve computation which is about $Q2^{n'd+n}$, especially when $d > 2$ (which is not uncommon).

Security Analysis

This novel MS-DSM scheme processes codewords, which transforms the representation of the shares. It is thus not directly amenable to automatic analysis, as is for instance enabled by [25]. Therefore, its verification poses a threat. We report in this chapter that our straightforward extension has a flaw, which is detected by carrying attacks. Namely, the number of shares is increased, and the number of traces to succeed the high-order attack increases, but then saturates at a given order (namely, 3 for words represented on 4 bits). This is subsequently analyzed under the prism of code properties: the dual distance of the masking material is indeed bounded to $4 = 3 + 1$ when the number of shares gets > 3.

A fix is proposed in Appendix 10.7, with an argument on the dual distance. The fix consists in using not the same H for all the masks m_w, $1 \leq w \leq d$, but in using d different codes H_w. Attacks on this revised scheme get strictly more complex (from the traces standpoint—more traces are needed to achieve a given success rate as the number of shares increases). While the attacks increasing complexity with the number of shares is not *per se* a proof for the new schemes, it nonetheless allows to validate its rationale.

Notice that, prior to our results, the study of the actual masking order was only possible using MI curves (slopes, see for instance, Fig. 5 from [26]). Now, it is possible to directly conduct the attacks.

$$Sum_q^{\mathsf{MS-DSM}}(z)$$
$$= \sum_{(m_1,\ldots,m_d) \in \Pi_{w=1}^d \mathbb{F}_2^{n'}} p(X_q^{(1)}|m_1) \ldots p(X_q^{(d)}|m_d) p(X_q^{(0)}|zG \oplus m_1 H_1 \cdots \oplus m_d H_d)$$
$$= \sum_{(m_1,\ldots,m_d) \in \Pi_{w=1}^d \mathbb{F}_2^{n'}} p'(X_q^{(1)}|m_1 H_1) \ldots p'(X_q^{(d)}|m_d H_d) p(X_q^{(0)}|zG \oplus m_1 H_1 \cdots \oplus m_d H_d)$$
$$= p'(X_q^{(1)}|.) \otimes \cdots \otimes p'(X_q^{(d)}|.) \otimes p_z(X_q^{(0)}|.)(0),$$

where: $p'(X|mH) = p(X|m)$ and $p_z(X_q^{(0)}|mH) = p(X_q^{(0)}|zG \oplus mH)$.

Subsequently, the $Sum_q^{\mathsf{MS-DSM}}$ becomes also a convolution relatively to the $\oplus$ law in the group $(\mathbb{F}_2^{n'}, \oplus)$. Thereby, it can be processed efficiently for each $z \in \mathbb{F}_2^k$, thanks to the WHT as follows:

$$Sum_q^{\mathsf{MS-DSM}}(z)$$
$$= WHT^{-1} \left(WHT(p'(X_q^{(1)}|.)) \bullet \cdots \bullet WHT(p'(X_q^{(d)}|.)) \bullet WHT(p_z(X_q^{(0)}|.)) \right)(0)$$
$$= \frac{1}{2^{n'/2}} \sum_{m \in \mathbb{F}_2^{n'}} \left(WHT(p'(X_q^{(1)}|.)) \bullet \cdots \bullet WHT(p'(X_q^{(d)}|.)) \bullet WHT(p_z(X_q^{(0)}|.)) \right)(m).$$

Similarly to above, one can process the product function

$$WHT(p'(X_q^{(1)}|.)) \bullet \cdots \bullet WHT(p'(X_q^{(d)}|.))$$

only once (for each q), since it is independent from z. Also, one can ignores the factor $\frac{1}{2^{n'/2}}$ since it is independent from k. Its overall complexity is about $Qdn2^n$

So, the overall computation complexity of the H-O Template Attack distinguisher against the so-called MS-DSM is slightly different from that expected above (Eq.(10.2)). It is about $Qdn'2^{n'} + Qn'2^{n+n'}$ instead of $Qdn2^n$. It is noticeable that this complexity is still better than that of the naïve computation which is about $Q2^{n'd+n}$, especially when $d > 2$ (which is not uncommon).

10.4.4 Type of Fourier Transform for the Polynomial DSM (PDSM) Scheme

A DSM scheme written over the space $\mathbb{F}_2[\alpha]/\langle P(\alpha)\rangle$, such that the degree of the irreducible polynomial P equals $n + n'$ (instead of its isomorphic space $\mathbb{F}_2^{n+n'}$), is called Polynomial Direct sum Masking (PDSM) [27]. The PDSM is more adapted for the AES masking as they use the same space $\mathbb{F}_2[\alpha]/\langle P(\alpha)\rangle$.

A similar extension of the $(1 + 1)$-shares PDSM to the so-called $(d + 1)$-shares PDSM is possible and it is of the same interest as that of DSM. Thereby, a similar spectral approach is possible to compute the optimal distinguisher, thanks to the corresponding FFT of the additive finite group of $\mathbb{F}_2[\alpha]/\langle P(\alpha)\rangle$.

10.4.5 Type of Fourier Transform for the Rotating S-boxes Masking (RSM) Scheme

As shown in Sect. 4.2.7, for a security/complexity compromise in the masking of the AES S-boxes, authors of [28] introduce the Rotating S-boxes Masking (RSM) [29, 30]. This scheme is the Boolean masking, where the masks are chosen uniformly from a code [24] of length $n = 8$ in $\mathbb{F}_2$, either

- $\mathbb{C}_0 = \{\texttt{0x00}\}$ (no masking),
- $\mathbb{C}_1 = \{\texttt{0x00}, \texttt{0xff}\}$,
- $\mathbb{C}_2$, a non-linear code of length 8, size 12 and its generating matrix denoted G, or
- $\mathbb{C}_3$, is a linear code of length 8 and dimension 4 (see paper [31] for more details).

The last case is interesting since there are sixteen masks (*i.e.,* 16 masked S-boxes' tables) [24,32,33]. For computing efficiently the Sum_q^{RSM} against this scheme, let us denote by G the generating matrix of $\mathbb{C}_3$. Let us also denote by $\overline{G}$ the generating matrix of a complementary subspace $\overline{\mathbb{C}_3}$ of $\mathbb{C}_3$.

Subsequently, each $z \in \mathbb{F}_2^8$ can be written as a sum $z_1 G \oplus z_2 \overline{G}$, where $z_1, z_2 \in \mathbb{F}_2^4$, and the $Sum_q^{\text{RSM}}(z)$ becomes

$$
\begin{aligned}
Sum_q^{\text{RSM}}(z) &= \sum_{m^{(1)},\ldots,m^{(d)}\in\mathbb{C}_3} p(X_q^{(1)}|m^{(1)})\ldots p(X_q^{(d)}|m^{(d)})p(X_q^{(0)}|z\oplus m^{(1)}\cdots\oplus m^{(d)}) \\
&= \sum_{m_1,\ldots,m_d\in\mathbb{F}_2^4} p(X_q^{(1)}|m_1 G)\ldots p(X_q^{(d)}|m_d G)p(X_q^{(0)}|z_2\overline{G}\oplus(z_1\oplus m_1\cdots\oplus m_d)G) \\
&= \sum_{m_1,\ldots,m_d\in\mathbb{F}_2^4} p'(X_q^{(1)}|m_1)\ldots p'(X_q^{(d)}|m_d)p'_{z_2}(X_q^{(0)}|z_1\oplus m_1\cdots\oplus m_d) \\
&= p'(X_q^{(1)}|.) \otimes \cdots \otimes p'(X_q^{(d)}|.) \otimes p'_{z_2}(X_q^{(0)}|.)(z_1),
\end{aligned}
$$

where $\Big(p'(X_q^{(w)}|m_w)=p(X_q^{(w)}|m_w G)\Big)_{w=1,\ldots d}$ and $p'_{z_2}(X_q^{(0)}|z_1\oplus m) = p(X_q^{(0)}|z_2\overline{G}\oplus (z_1\oplus m)G)$. Subsequently, the Sum_q^{RSM} is also a convolution relatively to the $\oplus$ law in the group $(\mathbb{F}_2^4, \oplus)$. Thereby, for each $z_2 \in \mathbb{F}_2^4$, one can efficiently process the 16 $Sum_{(q,z_2)}^{\mathsf{RSM}}(.) = Sum_q^{\mathsf{RSM}}(z_2\overline{G} \oplus .)$, thanks to the WHT as follows:

$$Sum_q^{\mathsf{RSM}}(z) = Sum_q^{\mathsf{RSM}}(z_1 G \oplus z_2\overline{G}) = Sum_{(q,z_2)}^{\mathsf{RSM}}(z_1)$$

$$= WHT^{-1}\left(WHT(p'(X_q^{(1)}|.)) \bullet \cdots \bullet WHT(p'(X_q^{(d)}|.)) \bullet WHT(p'_{z_2}(X_q^{(0)}|.))\right)(z_1).$$

That means, one needs to compute only 16 (*i.e.*, $2^{n/2}$) functions $Sum_{(q,z_2)}^{\mathsf{RSM}}(.)$ with an overall complexity of $\mathcal{O}(\frac{n}{2}d2^{n/2})$ for each one. Finally, the overall complexity of the attack is about $\mathcal{O}(\frac{n}{2}Qd2^n)$, *i.e.*, $2\times$ faster then that against the perfect masking.

10.4.6 Type of Fourier Transform for the Leakage Squeezing Masking (LSM) Scheme

As shown in Sect. 4.2.6, in this scheme, the shares are like for perfect masking (Boolean masking with all possible values for the masks) except some bijective functions $(F_w)_{w=1,\ldots,d}$ are applied to the masks [34,35], That leads to a better mixing of bits [24,36]. Since F_w is bijective for each w, so Sum_q^{LSM} becomes as following:

$$Sum_q^{\mathsf{LSM}}(z) =$$

$$\sum_{m^{(1)},\ldots,m^{(d)}\in G} p(X_q^{(1)}|m^{(1)})\ldots p(X_q^{(d)}|m^{(d)})p(X_q^{(0)}|z \oplus F_1(m^{(1)})\cdots \oplus F_d(m^{(d)}))$$

$$= \sum_{m^{(1)},\ldots,m^{(d)}\in G} p'(X_q^{(1)}|F_1(m^{(1)}))\ldots p'(X_q^{(d)}|F_d(m^{(d)}))p(X_q^{(0)}|z \oplus F_1(m^{(1)})\cdots \oplus F_d(m^{(d)}))$$

$$= \sum_{F_1(m^{(1)}),\ldots,F_d(m^{(d)})\in G} p'(X_q^{(1)}|F_1(m^{(1)}))\ldots p'(X_q^{(d)}|F_d(m^{(d)}))p(X_q^{(0)}|z \oplus F_1(m^{(1)})\cdots \oplus F_d(m^{(d)}))$$

$$= \sum_{m^{(1)},\ldots,m^{(d)}\in G} p'(X_q^{(1)}|m^{(1)})\ldots p'(X_q^{(d)}|m^{(d)})p(X_q^{(0)}|z \oplus m^{(1)} \cdots \oplus m^{(d)})$$

$$= p'(X_q^{(1)}|.) \otimes \cdots \otimes p'(X_q^{(d)}|.) \otimes p(X_q^{(0)}|.)(z),$$

where $\left(p'(X_q^{(w)}|F_w(m^{(w)})) = p(X_q^{(w)}|m^{(w)})\right)_{w=1,\ldots d}$.
    Thereby,

$$Sum_q^{\mathsf{LSM}}(z) = WHT^{-1}\left(WHT(p'(X_q^{(1)}|.)) \bullet \cdots \bullet WHT(p'(X_q^{(d)}|.)) \bullet WHT(p(X_q^{(0)}|.))\right)(z)$$

10.5 Experiments

10.5.1 Results of High-Order Attacks on Boolean Masking

Simulations are carried on waveforms represented as in Fig. 10.2. The WHT is computed using the open source library `fwht` [37].

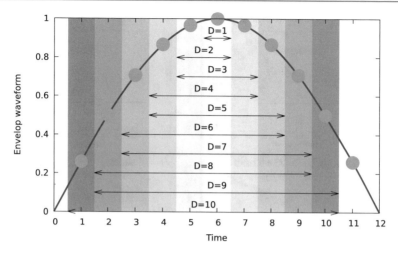

Fig. 10.2 Envelop of the $D = 1, 2, \ldots, 10$ considered waveforms

The attack results on windows of size $D = 1, 2, \ldots, 10$ are shown in Fig. 10.3, for several masking orders $(N = 1, \ldots, 8)$ in the case of Boolean masking. For a better comparison, the same attacks are also represented with the number of traces expressed in logarithmic scale, in Fig. 10.4.

Each of the graphs in Fig. 10.3 clearly demonstrate the usefulness to shift from monovariate HOOD $(D = 1)$ to multivariate HOOD $(D > 1)$. Indeed, the multivariate traces allow for attacks with fewer traces. Please notice that for $D = 1$, our new attack is the same as the HOOD, except that it is computed 32^d faster.

The computation gain of our Walsh-Hadamard approach is contrasted to that of the naïve approach in Fig. 10.8. The Figs. 10.3, 10.4 and 10.8 demonstrate the usefulness of the WHT, since the complexity is reduced from 2nd to $nd2^n$. This is a decrease from exponential to linear with respect to side-channel order d.

The attack on $n = 4$ bits (PRESENT S-Box) is represented in Fig. 10.5. One can see that, compared to AES $(n = 8$ bits), the attacks require a fewer number of traces.

10.5.2 Results of MS-DSM applied to PRESENT

In this section, we evaluate the security of the introduced "Multi-Share DSM" countermeasure applied on PRESENT $(n = 4$ bit). For the corresponding basic DSM, the selected codes $\mathbb{C}$ and $\mathscr{D}$ are generated by matrices G and H given below:

$$G = \begin{pmatrix} 1 & 0 & 0 & 0 & 0 & 0 & 1 & 1 \\ 0 & 1 & 0 & 0 & 1 & 0 & 1 & 0 \\ 0 & 0 & 1 & 0 & 1 & 1 & 0 & 0 \\ 0 & 0 & 0 & 1 & 0 & 1 & 0 & 1 \end{pmatrix} \quad \text{and} \quad H = \begin{pmatrix} 0 & 1 & 1 & 0 & 1 & 0 & 0 & 0 \\ 0 & 0 & 1 & 1 & 0 & 1 & 0 & 0 \\ 1 & 1 & 0 & 0 & 0 & 0 & 1 & 0 \\ 1 & 0 & 0 & 1 & 0 & 0 & 0 & 1 \end{pmatrix}.$$

Results on MS-DSM are shown in Fig. 10.6 for $\sigma = 2$. This figure shows that the incorrect construction (see Appendix 10.7.1) stops making attacks more complex, in terms of the number of traces to recover the key, when order increases (namely,

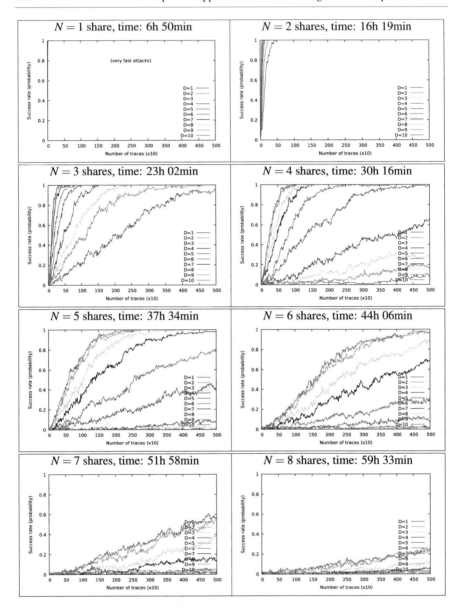

Fig. 10.3 Attacks on $D = 1, \ldots, 10$ samples on Boolean Masking of AES ($n = 8$ bit) for noise $\sigma = 1.0$, and various number of shares (for 100 attacks)

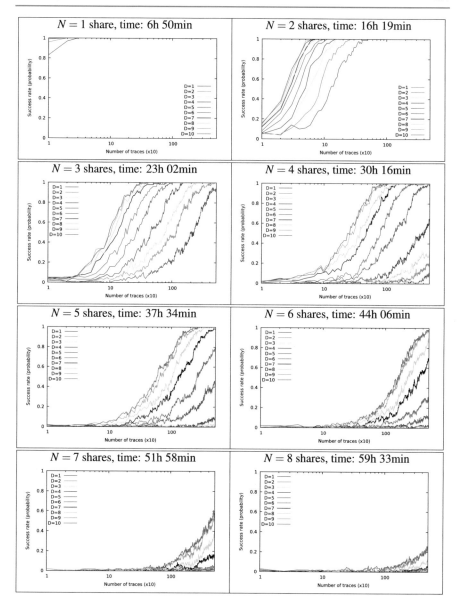

Fig. 10.4 Same as Fig. 10.3 (Boolean Masking on AES, $n = 8$ bit) but in log scale for the number of traces

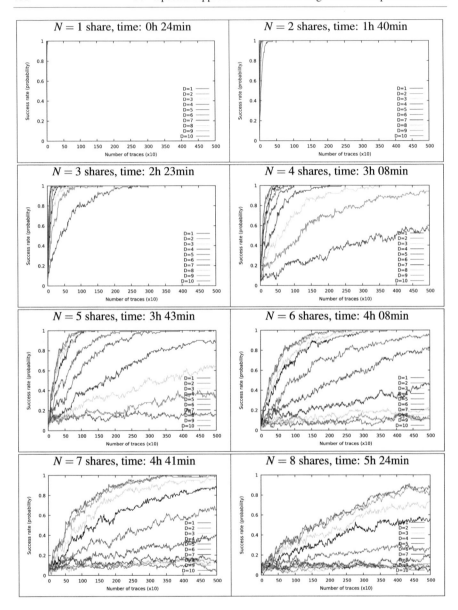

Fig. 10.5 Attacks on $D = 1, \ldots, 10$ samples on Boolean Masking of PRESENT ($n = 4$ bit) for noise $\sigma = 1.0$, and various number of shares (for 100 attacks)

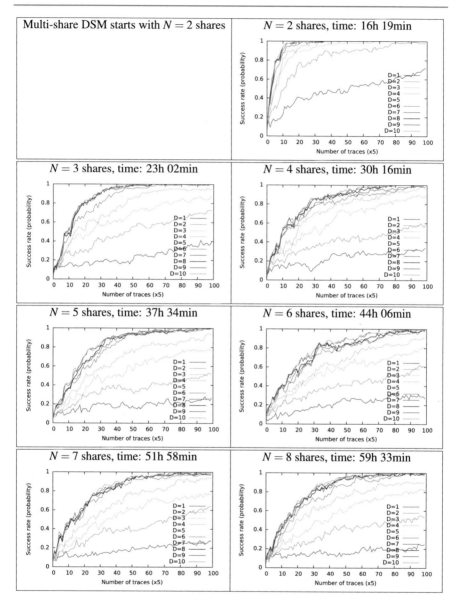

Fig. 10.6 Attacks on $D = 1, \ldots, 10$ samples for noise $\sigma = 2.0$, and various number of shares (for 100 attacks) against (incorrect, see Appendix 10.7.1) MS-DSM on PRESENT ($n = 4$ bit)

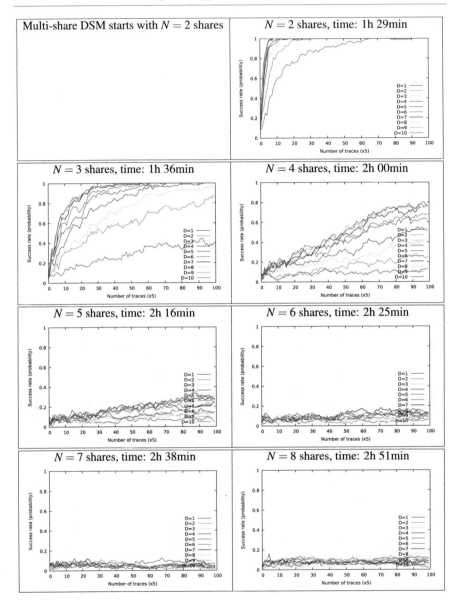

Fig. 10.7 Attacks on $D = 1, \ldots, 10$ samples for noise $\sigma = 2.0$, and various number of shares (for 100 attacks) against (correct, see Appendix 10.7.2) MS-DSM on PRESENT ($n = 4$ bit)

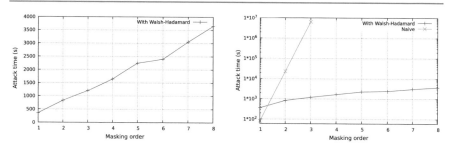

Fig. 10.8 Single attack time (on 5000 samples), as a function of its order (= number of shares)

there is no further improvement starting from order $d = 2$). Indeed, it is proven in the appendix that the security order is no more than $d_{\mathscr{D}}^{\perp} - 1 = 3 - 1$ (that is when $N = d + 1 = 3$). The behavior is reminiscent of the one already noted by Battistello et al. [38]: the attack countermeasure becomes no longer useful at a given noise level σ when the order increases. Such behavior could not have been seen by using the regular attack method, since the order is too high.

After fixing the MS-DSM (H_w are different), we get new results more secure (no flaw), as shown in Fig. 10.7.

10.5.3 Further Improvement in Performance

The attack time is almost linear as a function of the masking order, as shown in Fig. 10.8. Further performance improvement can be obtained by using multi-threaded code. According to the Algorithm 10.10, the bottlenecks of the attack computation are the lines 14, 14 and 15. Their complexities are, respectively, $\mathscr{O}(2^n \times Q \sum_w D^{(w)})$ (for the multivariate Gaussian noise case), $\mathscr{O}(n2^n \times Q(d+1))$, and $\mathscr{O}(2^n \times Q(d+1))$. Since the processing of the lines 14 and 14 are independent of a share to another, one can dispatch them to $d + 1$ threads, such that the wth thread processes the lines 14 and 14 for the corresponding wth share. For the 15th line, one can notice that the product can be computed out-of-order because the multiplication is commutative. This allows for further improvement by current concurrent product computation using a shared memory between the threads.

If the computer can run $d + 1$ threads in parallel, then the time to compute the attack is (almost) independent from the masking order.

10.6 Conclusion and Perspectives

10.6.1 Conclusion

Template attacks can be performed in the presence of masking. In this respect, the leakage from the different shares shall be profiled. The result is a collection of $d + 1$ distributions, characterized for each share value. For each share value, the profiling is a waveform (termed a sub-trace), and the attack naturally applies to this situation, even though the sub-traces have different sizes. We show in this chapter that the attack therefore consists of the convolution product of the $d+1$ distributions, independently of the masking scheme. The exact nature of the convolution depends on the type of masking (i.e., the random sharing method), and we show how to compute this convolution for six representative masking schemes.

The template attack online computation can be greatly speeded up by computing the convolution using spectral techniques (Plancherel equality). Furthermore, to still speed up the attack, we provide an interesting multi-threading implementation of our approach. In this respect, we show very high-order attacks on six flavors of masking schemes. In particular, we introduce a new masking scheme (multi-share extension of DSM) and show based on high-order attacks a third-order flaw. We then fix the scheme and validate it by attacks: the number of traces to succeed the attacks does increase as the number of shares increases.

10.6.2 Perspectives

The properties of the masking schemes highlighted in this chapter could benefit other studies related to side-channel analysis. For instance, a technique to formally verify masking schemes is presented in [25]. It consists of a symbolic verification that all tuples of shares which are supposed to be independent of the sensitive variable indeed are. The methodology bases itself upon simplification rules which are specific to Boolean masking, in that, it replaces expressions such as $X_1 \oplus E$, where E is an expression which depends on random variables $(X_i)_{i \in I}$, where $1 \notin I$, as R (which is randomly distributed). The method of Barthe [25] can be transported to other masking schemes by leveraging the transformations we put forward in this chapter.

10.7 Appendix: Multi-Share DSM Scheme

In this appendix, we explain how to extend DSM to a multi-share case. This appendix is divided into two subsections: how not to extend (Sect. 10.7.1) and how to it (Sect. 10.7.2). Both methods are supported by arguments based on coding theory.

10.7.1 Incorrect Multi-Share DSM Scheme

Recall About DSM

The DSM leverages two complementary codes $\mathbb{C}$ and $\mathscr{D}$ of generator matrix G and H. The unique share is denoted by $z^{(0)} = xG + mH$. As per Lemma 10.1, we are interested in the dual distance $d_{\mathscr{D}}^{\perp}$, because the bit-level security order is equal to $d_{\mathscr{D}}^{\perp} - 1$.

Extension of DSM: a First Attempt

The first generalization is that given in Eq. (10.10). Actually, although the sharing is now written with $(d + 1)$ shares, the representation can still be viewed as DSM, using

$$
\begin{cases}
\mathbf{G} = \begin{pmatrix} G & 0_{n \times (n+n')} & 0_{n \times (n+n')} & \cdots & 0_{n \times (n+n')} \end{pmatrix} & \text{in lieu of } G, \text{ and} \\[2mm]
\mathbf{H} = \begin{pmatrix}
H & I_{n'} & 0_{n' \times n'} & \cdots & 0_{n \times n'} \\
H & 0_{n' \times n'} & I_{n'} & \cdots & 0_{n \times n'} \\
\vdots & 0_{n' \times n'} & 0_{n' \times n'} & \ddots & \\
H & 0_{n' \times n'} & 0_{n \times n'} & \cdots & I_{n'}
\end{pmatrix} & \text{in lieu of } H.
\end{cases} \tag{10.11}
$$

It is easy to see that the dual of spacevector generated by $\mathbf{H}$ is generated by

$$
\begin{pmatrix} I_{n+n'} & H^T & \cdots & H^T \end{pmatrix}. \tag{10.12}
$$

This means that all rows of Eq. (10.12) are orthogonal with all rows of $\mathbf{H}$ as defined in Eq. (10.11). The security of this new masking scheme is thus equal to $d^{\perp}_{\mathrm{span}(\mathbf{H})} - 1$. It is equal to the minimum distance of the code spawn by the matrix represented in Eq. (10.12) (minus the number one).

It could be believed that the minimum distance of the matrix represented in Eq. (10.12) grows when the number of H^T blocks increases. This is not true, since the minimum distance can saturate. This happens when there exists a linear combination of lines in H^T that is equal to the null vector. Typically, we used

$$
H = \begin{pmatrix}
0 & 1 & 1 & 0 & 1 & 0 & 0 & 0 \\
0 & 0 & 1 & 1 & 0 & 1 & 0 & 0 \\
1 & 1 & 0 & 0 & 0 & 0 & 1 & 0 \\
1 & 0 & 0 & 1 & 0 & 0 & 0 & 1
\end{pmatrix}
$$

Clearly, the sum of the four first columns is equal to the all-zero column, hence after transposition, the sum of the four first lines of H^T is equal to zero.

So, at most, the minimum dual distance of the code $\mathbf{H}$ is 5, and the maximum security level is 4.

This explains why the attacks do not become more difficult as d increases (in Fig. 10.6).

10.7.2 Correct Multi-Share DSM Scheme

Proposed Correction, Inspired from IPM

The limitation identified in the first attempt to make MS-DSM came from the fact all masks $m_1, \ldots, m_d$, where processed by the same code $\mathscr{D}$. One wishes to replace Eq. (10.12) by

$$
\begin{pmatrix} I_{n+n'} & H_1^T & \cdots & H_d^T \end{pmatrix}. \tag{10.13}
$$

It is then possible have a growth of the minimum distance of the spacevector spawn of this matrix linearly with d. Thus, instead of Eq. (10.11), the correct MS-DSM scheme therefore employs

$$
\begin{cases}
\mathbf{G} = \begin{pmatrix} G & 0_{n \times (n+n')} & 0_{n \times (n+n')} & \cdots & 0_{n \times (n+n')} \end{pmatrix} & \text{in lieu of } G, \text{ and} \\[2ex]
\mathbf{H} = \begin{pmatrix}
H_1 & I_{n'} & 0_{n' \times n'} & \cdots & 0_{n \times n'} \\
H_2 & 0_{n' \times n'} & I_{n'} & \cdots & 0_{n \times n'} \\
\vdots & 0_{n' \times n'} & 0_{n' \times n'} & \ddots & \\
H_d & 0_{n' \times n'} & 0_{n \times n'} & \cdots & I_{n'}
\end{pmatrix} & \text{in lieu of } H.
\end{cases}
\tag{10.14}
$$

Validation

This new scheme works in practice, as attested by the attacks requiring more and more traces to reach a given success rate, (as show in Fig. 10.7).

References

1. Oswald E, Mangard S (2007) Template attacks on masking — resistance is futile. In: Abe M (ed), CT-RSA. Lecture notes in computer science, vol 4377. Springer, pp 243–256
2. Chari S, Rao JR, Rohatgi P (2002) Template attacks. In: Kaliski BS Jr, Koç ÇK, Paar C (eds), Cryptographic hardware and embedded systems - CHES 2002, 4th international workshop, redwood shores, CA, USA, August 13-15, 2002, Revised papers. Lecture notes in computer science. Springer, pp 13–28
3. Lomné V, Prouff E, Rivain M, Roche T, Thillard A, How to estimate the success rate of higher-order side-channel attacks. In: Batina, Robshaw (eds), [14], pp 35–54
4. Bruneau N, Guilley S, Heuser A, Rioul O (2014) Masks will fall off – higher-order optimal distinguishers. In: Sarkar P, Iwata T (eds), Advances in cryptology – ASIACRYPT 2014 - 20th international conference on the theory and application of cryptology and information security, Kaoshiung, Taiwan, R.O.C., December 7–11, 2014, Proceedings, Part II. Lecture notes in computer science, vol 8874. Springer, pp 344–365
5. Lemke-Rust K, Paar C (2007) Gaussian mixture models for higher-order side channel analysis. In: Paillier P, Verbauwhede I (eds), CHES. LNCS, vol 4727. Springer, pp 14–27
6. Bruneau N, Guilley S, Heuser A, Marion D, Rioul O (2017) Optimal side-channel attacks for multivariate leakages and multiple models. J Cryptograph Eng 7(4):331–341
7. Bruneau N, Guilley S, Heuser A, Rioul O, Standaert F-X, Teglia Y (2016) Taylor expansion of maximum likelihood attacks for masked and shuffled implementations. In: Cheon JH, Takagi T (eds), Advances in cryptology - ASIACRYPT 2016 - 22nd international conference on the theory and application of cryptology and information security, Hanoi, Vietnam, December 4–8, 2016, Proceedings, Part I. Lecture notes in computer science, vol 10031, pp 573–601
8. Ouladj M, El Mrabet N, Guilley S, Guillot P, Millérioux G (2020) On the power of template attacks in highly multivariate context. J Cryptogr Eng 10(4):337–354
9. Prouff E, Rivain M, Bevan R (2009) Statistical analysis of second order differential power analysis. IEEE Trans Comput 58(6):799–811
10. Standaert F-X, Veyrat-Charvillon N, Oswald E, Gierlichs B, Medwed M, Kasper M, Mangard S (2010) The world is not enough: another look on second-order DPA. In: ASIACRYPT. LNCS, vol 6477. Springer, Singapore, pp 112–129. http://www.dice.ucl.ac.be/~fstandae/PUBLIS/88.pdf
11. Chari S, Jutla CS, Rao JR, Rohatgi P (1999) Towards sound approaches to counteract power-analysis attacks. In: Wiener MJ (ed), CRYPTO. Lecture notes in computer science, vol 1666. Springer, pp 398–412

12. Prouff E, Rivain M (2013) Masking against side-channel attacks: a formal security proof. In: Johansson T, Nguyen PQ (eds), Advances in cryptology - EUROCRYPT 2013, 32nd annual international conference on the theory and applications of cryptographic techniques, Athens, Greece, May 26–30, 2013. Proceedings. Lecture notes in computer science, vol 7881. Springer, pp 142–159

13. Massey JL (1992) Linear codes with complementary duals. Discret Math 106–107:337–342

14. Ngo XT, Bhasin S, Danger J-L, Guilley S, Najm Z (2015) Linear complementary dual code improvement to strengthen encoded circuit against hardware Trojan horses. In: IEEE international symposium on hardware oriented security and trust, HOST 2015, Washington, DC, USA, 5–7 May, 2015. IEEE, pp 82–87

15. Goubin L (2001) A sound method for switching between Boolean and arithmetic masking. In: Koç ÇK, Naccache D, Paar C (eds), CHES. Lecture notes in computer science, vol 2162. Springer, pp 3–15

16. Bruneau N, Guilley S, Heuser A, Marion D, Rioul O, Less is more - dimensionality reduction from a theoretical perspective. In: Güneysu, Handschuh [99], pp 22–41

17. Kerstin L, Kai S, Paar C (2004) DPA, on n-bit sized Boolean and arithmetic operations and its application to IDEA, RC6, and the HMAC-construction. In: CHES. Lecture notes in computer science, vol 3156. Springer, Cambridge, pp 205–219

18. Dziembowski S, Faust S (2011) Leakage-resilient cryptography from the inner-product extractor. In: Lee DH, Wang X (eds), Advances in cryptology - ASIACRYPT 2011 - 17th international conference on the theory and application of cryptology and information security, Seoul, South Korea, December 4–8, 2011. Proceedings. Lecture notes in computer science, vol 7073. Springer, pp 702–721

19. Balasch J, Faust S, Gierlichs B, Verbauwhede I (2012) Theory and practice of a leakage resilient masking scheme. In: Wang X, Sako K (eds), Advances in cryptology - ASIACRYPT 2012 - 18th international conference on the theory and application of cryptology and information security, Beijing, China, December 2–6, 2012. Proceedings. Lecture notes in computer science, vol 7658. Springer, pp 758–775

20. Balasch J, Faust S, Gierlichs B, Inner product masking revisited. In: Oswald, Fischlin (eds), [154], pp 486–510

21. Balasch J, Faust S, Gierlichs B, Paglialonga C, Standaert F-X (2017) Consolidating inner product masking. In: Takagi T, Peyrin T (eds), Advances in cryptology - ASIACRYPT 2017 - 23rd international conference on the theory and applications of cryptology and information security, Hong Kong, China, December 3–7, 2017, Proceedings, Part I. Lecture notes in computer science, vol 10624. Springer, pp 724–754

22. Poussier R, Guo Q, Standaert F-X, Carlet C, Guilley S (2017) Connecting and improving direct sum masking and inner product masking. In: Eisenbarth T, Teglia Y (eds) Smart card research and advanced applications - 16th international conference, CARDIS 2017, Lugano, Switzerland, November 13–15, 2017, Revised selected papers. Lecture notes in computer science, vol 10728. Springer, pp 123–141

23. Bringer J, Carlet C, Chabanne H, Guilley S, Maghrebi H (2014) Orthogonal direct sum masking – a smartcard friendly computation paradigm in a code, with builtin protection against side-channel and fault attacks. In: WISTP. LNCS, vol 8501. Springer, Heraklion, pp 40–56

24. Guilley S, Heuser A, Rioul O, Codes for side-channel attacks and protections. In: Hajji et al. [100], pp 35–55

25. Barthe G, Belaïd S, Dupressoir F, Fouque P-A, Grégoire B, Strub P-Y, Verified proofs of higher-order masking. In: Oswald, Fischlin [154], pp 457–485

26. Carlet C, Danger J-L, Guilley S, Maghrebi H, Prouff E (2014) Achieving side-channel high-order correlation immunity with leakage squeezing. J. Cryptograph Eng 4(2):107–121

27. Cedric T, Carlet C, Guilley S, Daif A (2018) Polynomial direct sum masking to protect against both sca and fia. J Cryptograph Eng

28. Nassar M, Souissi Y, Guilley S, Danger J-L (2012) RSM: a small and fast countermeasure for AES, secure against 1st and 2nd-order zero-offset SCAs. In: Rosenstiel W, Thiele L (eds), 2012 Design, automation and test in Europe conference and exhibition, DATE 2012, Dresden, Germany, March 12-16, 2012. IEEE, pp 1173–1178

29. Kutzner S, Poschmann A, On the security of RSM - presenting 5 first- and second-order attacks. In: Prouff [172], pp 299–312

30. Yamashita N, Minematsu K, Okamura T, Tsunoo Y (2014) A smaller and faster variant of RSM. In: DATE. IEEE, pp 1–6

31. Carlet C, Guilley S (2013) Side-channel indistinguishability. In: HASP. ACM, New York, pp 9:1–9:8

32. Carlet C, Correlation-immune boolean functions for leakage squeezing and rotating S-box masking against side channel attacks. In: Gierlichs et al. [85], pp 70–74

33. DeTrano A, Karimi N, Karri R, Guo X, Carlet C, Guilley S (2015) Exploiting small leakages in masks to turn a second-order attack into a first-order attack and improved rotating substitution box masking with linear code cosets. Sci World J, 10. https://doi.org/10.1155/2015/743618

34. Carlet C, Danger J-L, Guilley S, Maghrebi H (2012) Leakage squeezing of order two. In: Galbraith SD, Nandi M (eds) Progress in cryptology - INDOCRYPT 2012, 13th international conference on cryptology in India, Kolkata, India, December 9-12, 2012. Proceedings. Lecture notes in computer science, vol 7668. Springer, pp 120–139

35. Karmakar S, Chowdhury DR (2013) Leakage squeezing using cellular automata. In: Kari J, Kutrib M, Malcher A (eds) Automata. Lecture notes in computer science, vol 8155. Springer, pp 98–109

36. Carlet C, Danger J-L, Guilley S, Maghrebi H (2014) Leakage squeezing: optimal implementation and security evaluation. J Math Cryptol 8(3):249–295

37. Luo D (2015) `fwht`: fast walsh hadamard transform in Python. https://github.com/dingluo/fwht

38. Battistello A, Coron J-S, Prouff E, Zeitoun R (2016) Horizontal side-channel attacks and countermeasures on the ISW masking scheme. In: Gierlichs B, Poschmann AY (eds), Cryptographic hardware and embedded systems - CHES 2016 - 18th international conference, Santa Barbara, CA, USA, August 17–19, 2016, Proceedings. Lecture notes in computer science, vol 9813. Springer, pp 23–39

Printed in the United States
by Baker & Taylor Publisher Services